SURVIVING
SUPPLY CHAIN
INTEGRATION

Strategies for
Small Manufacturers

Committee on Supply Chain Integration
Board on Manufacturing and Engineering Design
Commission on Engineering and Technical Systems
National Research Council

NATIONAL ACADEMY PRESS
Washington, D.C.

National Academy Press • 2101 Constitution Avenue, N.W. • Washington, DC 20418

NOTICE: The project that is the subject of this report was approved by the Governing Board of the National Research Council, whose members are drawn from the councils of the National Academy of Sciences, the National Academy of Engineering, and the Institute of Medicine. The members of the committee responsible for the report were chosen for their special competencies and with regard for appropriate balance.

This study by the Board on Manufacturing and Engineering Design was conducted under MURC Grant No. 111-94-0007-00 from the Robert C. Byrd Institute and the National Institute of Standards and Technology. Any opinions, findings, conclusions, or recommendations expressed in this publication are those of the author(s) and do not necessarily reflect the views of the Robert C. Byrd Institute and National Institute of Standards and Technology.

Library of Congress Cataloging-in-Publication Data

Surviving supply chain integration : challenges for small manufacturers / Committee on Supply Chain Integration, Board on Manufacturing and Engineering Design, Commission on Engineering and Technical Systems, National Research Council.
    p. cm.
Includes biliographical references and index.
    ISBN 0-309-06878-9 (casebound)
    1. Business logistics. 2. Small business—Management. I. National Research Council (U.S.). Committee on Supply Chain Integration. II. Title.
    HD38.5 .S897 2000
    670'.68—dc21

00-008199

Surviving Supply Chain Integration: Strategies for Small Manufacturers is available from the National Academy Press, 2101 Constitution Ave., N.W, Lockbox 285, Washington,DC 20055; (800) 624-6242 or (202) 334-3313 (in the Washington metropolitan area); Internet <http://www.nap.edu>.

Printed in the United States of America.

THE NATIONAL ACADEMIES

National Academy of Sciences
National Academy of Engineering
Institute of Medicine
National Research Council

The **National Academy of Sciences** is a private, nonprofit, self-perpetuating society of distinguished scholars engaged in scientific and engineering research, dedicated to the furtherance of science and technology and to their use for the general welfare. Upon the authority of the charter granted to it by the Congress in 1863, the Academy has a mandate that requires it to advise the federal government on scientific and technical matters. Dr. Bruce M. Alberts is president of the National Academy of Sciences.

The **National Academy of Engineering** was established in 1964, under the charter of the National Academy of Sciences, as a parallel organization of outstanding engineers. It is autonomous in its administration and in the selection of its members, sharing with the National Academy of Sciences the responsibility for advising the federal government. The National Academy of Engineering also sponsors engineering programs aimed at meeting national needs, encourages education and research, and recognizes the superior achievements of engineers. Dr. William A. Wulf is president of the National Academy of Engineering.

The **Institute of Medicine** was established in 1970 by the National Academy of Sciences to secure the services of eminent members of appropriate professions in the examination of policy matters pertaining to the health of the public. The Institute acts under the responsibility given to the National Academy of Sciences by its congressional charter to be an adviser to the federal government and, upon its own initiative, to identify issues of medical care, research, and education. Dr. Kenneth I. Shine is president of the Institute of Medicine.

The **National Research Council** was organized by the National Academy of Sciences in 1916 to associate the broad community of science and technology with the Academy's purposes of furthering knowledge and advising the federal government. Functioning in accordance with general policies determined by the Academy, the Council has become the principal operating agency of both the National Academy of Sciences and the National Academy of Engineering in providing services to the government, the public, and the scientific and engineering communities. The Council is administered jointly by both Academies and the Institute of Medicine. Dr. Bruce M. Alberts and Dr. William A. Wulf are chairman and vice chairman, respectively, of the National Research Council.

COMMITTEE ON SUPPLY CHAIN INTEGRATION

JAMES LARDNER (chair), Deere & Company (retired), Davenport, Iowa
STEVEN J. BOMBA, Johnson Controls, Inc., Glendale, Wisconsin
JOHN A. CLENDENIN, Harvard Business School, Cambridge,
   Massachusetts
GERALD E. JENKS, The Boeing Company, Chesterfield, Missouri
JACK J. KLIM, JR., D&E Industries, Huntington, West Virginia
EDWARD KWIATKOWSKI, Supply America Corporation, Chagrin
   Falls, Ohio
HAU LEE, Stanford University, Palo Alto, California
CHARLES W. LILLIE, Science Applications International Corporation,
   McLean, Virginia
MARY C. MURPHY-HOYE, Intel Corporation, Chandler, Arizona
JAMES R. MYERS, Kilpatrick Stockton LLP, Washington, D.C.
JAMES B. RICE, JR., Massachusetts Institute of Technology, Cambridge
OLIVER WILLIAMSON, University of California, Berkeley
THOMAS YOUNG, Lockheed Martin Corporation (retired), Potomac,
   Maryland

Staff of the Board on Manufacturing and Engineering Design

ROBERT RUSNAK, Senior Program Officer (until October 1998)
JOHN F. RASMUSSEN, Senior Program Officer (since November 1998)
THOMAS E. MUNNS, Associate Director
AIDA C. NEEL, Senior Project Assistant
TERI THOROWGOOD, Research Associate

Liaison with the Board on Manufacturing and Engineering Design

FRIEDRICH B. PRINZ, Stanford University, Palo Alto, California

Liaison Representatives

BRAD BOTWIN, U.S. Department of Commerce, Washington, D.C.
KEVIN CARR, National Institute of Standards and Technology,
   Gaithersburg, Maryland
PHIL NANZETTA, Strategic Focus, Rockville, Maryland
MARIA STOPHER, National Institute of Standards and Technology,
   Gaithersburg, Maryland
STEVEN WAX, Defense Advanced Research Projects Agency,
   Arlington, Virginia
CHARLOTTE WEBER, Robert C. Byrd Institute, Huntington, West
   Virginia

v

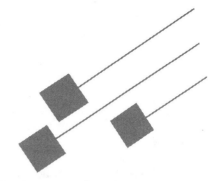

Acknowledgments

The Committee on Supply Chain Integration would like to thank the following individuals for their presentations: P. Jeffrey Trimmer, DaimlerChrysler; Frederic E. Rakness, Lockheed Martin; Susan Moehring, Institute of Advanced Manufacturing Sciences, Inc.; Dale Crownover, Texas Nameplate Company, Inc.; David Salazar, General Technology Corporation; Troy Takach, The Parvus Corporation; Robert Squier, Curtis Screw Company; and Charlotte Weber, the Robert C. Byrd Institute. The committee would also like to thank the representatives of the small and medium-sized enterprises who participated in the survey and the field agents of the Manufacturing Extension Partnership and the Robert C. Byrd Institute who administered it.

This report has been reviewed by individuals chosen for their diverse perspectives and technical expertise, in accordance with procedures approved by the NRC's Report Review Committee. The purpose of this independent review is to provide candid and critical comments that will assist the authors and the NRC in making the published report as sound as possible and to ensure that the report meets institutional standards for objectivity, evidence, and responsiveness to the study charge. The contents of the review comments and draft manuscript remain confidential to protect the integrity of the deliberative process. We wish to the thank the following individuals for their participation in the review of this report: Bruce Blagg, Transformingit; Morris A. Cohen, University of Pennsylvania; Robert W. Hall, Indiana University; Robert B. Handfield, North Carolina State University; Bernard LaLond, Ohio State University; Terrance

Pohlen, University of North Florida; Joel Samuel Yudken, AFL-CIO, and Mohamad Zarrugh, James Madison University.

While the individuals listed above have provided many constructive comments and suggestions, responsibility for the final content of the report rests solely with the authoring committee and the NRC.

Finally, the committee gratefully acknowledges the support of the staff of the Board on Manufacturing and Engineering Design, including Robert Rusnak, study director (until March 1998); Thomas E. Munns, study director (until October 1998); John F. Rasmussen, study director (since November 1998); Teri Thorowgood, research associate; and Aida C. Neel, senior project assistant. The report was edited by Carol R. Arenberg, Commission on Engineering and Technical Systems

Preface

In the early 1980s, it became apparent to many that manufacturing industries in the United States were losing their ability to compete in world markets. The erosion of domestic market share was particularly alarming in industries that had been the exclusive province of U.S. companies, including automobiles, machine tools, and electronics.

Concerns about this situation led several government agencies and departments, among them the U.S. Department of Defense and the National Science Foundation, to ask the National Research Council to examine the problem and recommend solutions. Between 1986 and 1994, the Manufacturing Studies Board of the National Research Council undertook several studies in which they identified fundamental deficiencies in the way U.S. manufacturers addressed the issues of cost, quality, and time to market. One aspect of the problem that did not command sufficient attention at the time was the long-held belief on the part of U.S. manufacturers that the integration of manufacturing operations, both vertical and horizontal, always provides a competitive advantage.

As manufacturers responded to these market challenges and learned more about their foreign competitors, it became increasingly apparent that too much integration could be a disadvantage. Therefore, many U.S. manufacturers began to focus investments and attention on honing their "core competencies" while procuring the rest of the goods and services required to produce their end products from others. This change in strategy increased their dependency on their suppliers and expanded the challenge of managing a diverse agglomeration of direct suppliers and suppliers to suppliers.

The range of products and services provided by these suppliers has become very large, making management of supply chains increasingly complex. This causes a variety of problems, not only for original equipment manufacturers (OEMs) and prime government contractors (the end-product producers), but also for other participants in these supply systems. With increasing market pressure to shorten product development cycles, reduce costs, and improve quality, suppliers too are facing more demanding managerial and operational requirements. Meeting these requirements can be especially challenging for small and medium-sized manufacturing enterprises (SMEs).

Faced with these fundamental changes in the role of SMEs in manufacturing supply chains, the National Institute of Standards and Technology (NIST) and the Robert C. Byrd Institute (RCBI) requested that the National Research Council (NRC) identify the new, more demanding requirements for supply chain participation and recommend ways that SMEs could be assisted in addressing them.

NIST oversees the Manufacturing Extension Partnership, a nationwide program that advises and assists manufacturing businesses with 500 employees or less on issues that affect their competitiveness in the changing manufacturing environment. RCBI is a national program whose mission is to create a quality supplier base for the U.S. Department of Defense and its prime contractors through "teaching factories," computer integration, and workforce development. Both organizations recognize that competent, competitive suppliers operating in efficient, modern supply chains are essential to the competitiveness of U.S. end-product manufacturers in world markets.

In response to their request, the NRC established the Committee on Supply Chain Integration under the direction of the Board on Manufacturing and Engineering Design. To enhance the committee's understanding of SMEs, a survey was conducted of randomly selected SMEs from the NIST Manufacturing Extension Partnership database. In addition, a number of small, successful manufacturing suppliers were invited to meet with the committee for a firsthand exchange of ideas about the challenges and problems of participating in the integrated supply chains of large OEMs.

The committee found that, although there is great diversity in U.S. manufacturing, succesful SMEs possess a number of common capabilities. Nevertheless, the committee emphasizes that each SME must carefully assess its own circumstances in the rapidly changing business environment, identify gaps between supply chain requirements and its own capabilities, and find ways to fill the gaps. The committee's recommendations are based on the assumption that the focus on core competencies and outsourcing trends will continue for the foreseeable future and that

U.S. industry will follow the integrated supply chain model in its drive to remain competitive in the increasingly global economy.

This report is not intended to be a definitive text on supply chain integration. Rather, it attempts to identify the converging effects of supply chain integration and changing technologies on SMEs and to recommend to SMEs and the manufacturing extension centers and technical resource providers that support them specific approaches for dealing with these issues. Some of the recommendations may seem very basic, but they are included because many SMEs have yet to take the basic steps essential for their survival.

Comments on this report can be sent by electronic mail to *bmaed@nas.edu* or by fax to BMAED (202) 334-3718.

James Lardner, *chair*
Committee on Supply Chain Integration

Contents

PART II
SMALL AND MEDIUM-SIZED MANUFACTURING ENTERPRISES
IN INTEGRATED SUPPLY CHAINS

Tables and Figures

Executive Summary

The business environment is changing rapidly and with it the supply chains of original equipment manufacturers (OEMs). Most OEMs no longer compete solely as autonomous corporations. They also compete as participants in integrated supply chains. In response to competitive pressures, U.S. manufacturers are purchasing increasing amounts of goods and services from outside suppliers (i.e., outsourcing), as well as integrating their supply chains to improve performance. A 1998 survey revealed that 80 percent of manufacturers had formal supply chain management programs or planned to start them in the next year. In this context, the National Institute of Standards and Technologies and the Robert C. Byrd Institute requested that the National Research Council conduct a study of the new roles and challenges faced by small and medium-sized manufacturing enterprises (SMEs) in integrated supply chains. The Committee on Supply Chain Integration was formed, under the auspices of the Board on Manufacturing and Engineering Design, for this purpose.

STATEMENT OF TASK

The committee was asked to perform the following specific tasks:

- Identify and analyze state-of-the-art supply chain integration concepts.
- Define the requirements for successful participation by SMEs in integrated supply chains.

1

- Identify the gaps between integrated supply chain requirements and the capabilities of SMEs.
- Suggest strategies to assist SMEs in developing the capabilities required for successful participation.

This report is intended for the owners and managers of small and medium-sized manufacturing enterprises and for the manufacturing extension centers and technical resource providers (MEC/TRPs) that support them.

SMALL AND MEDIUM-SIZED
MANUFACTURING ENTERPRISES

The estimated 330,000 SMEs in the United States have a substantial economic impact. Defined as having fewer than 500 employees, SMEs are important to the nation because they account for 98 percent of all manufacturing plants, employ two-thirds of the nation's 18 million manufacturing workers, generate more than half of the total value added in the manufacturing sector, and are the source of many innovations in technology.

SMEs typically provide capabilities that their larger customers do not have or cannot cost-effectively create, such as:

- agility in responding to changes in technologies, markets, and trends
- efficiency due, in part, to less bureaucracy
- initiative and entrepreneurial behavior on the part of employees resulting in higher levels of creativity and energy and a greater desire for success
- access to specialized proprietary technologies, process capabilities, and expertise
- shorter time-to-market because operations are small and focused
- lower labor costs and less restrictive labor contracts
- spreading the costs of specialized capabilities over larger production volumes by serving multiple customers
- lower cost, customer focused, and customized services, including documentation, after-sales support, spare parts, recycling, and disposal

SUPPLY CHAINS

The committee defined a *supply chain* as an association of customers and suppliers who, working together yet in their own best interests, buy, convert, distribute, and sell goods and services among themselves resulting in the creation of a specific end product. A supply chain includes all of the capabilities and functions required to design, fabricate, distribute, sell, support, use, and recycle or dispose of a product. An *integrated supply chain* can be defined as an association of customers and suppliers who work together to optimize their collective performance in the creation, distribution, and support of an end product. The objective of integration is to focus and coordinate the relevant resources of each participant on the needs of the supply chain and to optimize the overall performance of the chain.

GENERAL REQUIREMENTS AND CAPABILITIES

SMEs cannot ignore the supply chain revolution and remain competitive. Although the outsourcing trend is providing increased opportunities for suppliers, trends toward globalization and increased supply chain integration pose serious challenges. A comparison of the requirements of each OEM's supply chain with the unique capabilities of an SME and its own supply chains often reveals deficiencies or gaps the SME must address. Closing these gaps involves (1) eliminating or circumventing constraints, (2) obtaining appropriate capabilities to satisfy the unmet need, and (3) utilizing the capabilities in a timely and effective manner. Although each SME has its own definition of success, the committee defined *successful supply chain participation* in a manner similar to the traditional definition of *business success* (i.e., participation in which benefits to the participant substantially exceed the costs).

REQUIREMENTS, CAPABILITIES, AND GAPS

Competitive cost, quality, service, and delivery are the traditional fundamental capabilities required of all suppliers, but successful participation in today's integrated supply chains requires more. Although these evolving fundamental capabilities are still the cornerstones of supply chain requirements, technology and management skills are increasingly critical for success.

Quality

Today's global markets demand products of higher quality, but high-quality products cannot be assembled cost effectively from low-quality components. Therefore, to remain competitive, OEMs are demanding higher quality products from their suppliers. Suppliers with quality deficiencies weaken the entire supply chain and, unless they improve, are being phased out. Quality is even more critical in supply chains using just-in-time manufacturing and low inventory levels because there are fewer buffers to protect against quality failures.

Thus, SMEs should not consider quality as only a requirement for continued supply chain participation, but as a strategic capability. SMEs that adopt quality as a competitive strategy find that they are better able to weather cyclical swings in their business and that their product costs are lower.

Recommendation. In response to the requirements of integrated supply chains for improved quality, small and medium-sized manufacturing enterprises should adopt quality as a competitive strategy and consider implementing techniques, such as six sigma, ISO certification, and statistical process controls, to comply with customer demands, improve overall business performance, and provide a common language for communication on quality issues.

Cost and Value

Costs have always been critical, and in the increasingly global economy it is not unusual for SMEs to find sudden gaps between their prices and the prices of competitors from low-cost areas. The convergence of (1) improvements in high-speed communications, (2) reductions in transportation costs, (3) the widespread adoption of English as the language of business, and (4) universal access to technology and effective management practices has enabled companies in areas with low labor costs to become competitive regardless of their location. Thus, SMEs must substantially reduce costs by using both traditional and innovative approaches, such as integrating their own supply chains.

Many SMEs will have to do more than provide low-cost parts if they want to become partners with demanding customers. They will also have to provide value-added services, such as low-cost storage, rapid response to warranty issues, ready access to spare parts, improved logistics, and increased design capabilities. Because their present customers may be unwilling to pay for added value, SMEs may have to reposition themselves into new industries and find customers that are willing to pay for value-added products and services.

Recommendation. Small and medium-sized manufacturing enterprises should rigorously reduce costs internally and throughout their supply chains. They should also seek ways to increase the value added to their products and services and find customers that are willing to reward such value.

Delivery

With increased levels of supply chain integration and reduced inventory levels, reliable, on-time deliveries have become critical for success. Large inventories and production capacities were traditionally required to ensure on-time delivery. However, with advanced information systems, agile manufacturing organizations with flexible equipment and tooling, and sophisticated logistics systems, integrated supply chains no longer need large, costly inventory buffers to respond to unexpected events and variations in demand. These capabilities should be augmented by the effective use of advanced transportation capabilities, such as overnight delivery.

Recommendation. In response to increasing demands for rapid delivery and customized products, small and medium-sized manufacturing enterprises should consider using advanced supply chain communication systems, flexible manufacturing techniques, and modern transportation capabilities as alternatives to investing in large inventories and production capacities.

Service

Customer expectations for timely service before, during, and after a sale continue to increase and, aided by the Internet and modern transportation methods, suppliers are responding to these demands. Web sites are being used to post maintenance manuals, service bulletins, and responses to frequently asked questions. e-Commerce enables customers to place orders around the clock from anywhere in the world without incurring the cost of long-distance calls. Replacement parts can be delivered overnight in the United States and within a few days in most of the rest of the world.

Recommendation. In response to increasing customer expectations for service and support, small and medium-sized manufacturing enterprises (SMEs) should reassess their service and support capabilities and revise them, as needed, to remain competitive and to seize new market opportunities. SMEs should develop an understanding of the opportunities provided by various Web technologies and, if appropriate, create a Web presence.

Building Partnerships

Partnerships are the backbone of integrated supply chains. In many cases, partnerships provide opportunities for competitive advantage at lower cost than vertical integration.

Recommendation. Small and medium-sized manufacturing enterprises should develop a basic understanding of partnership agreements and, with legal assistance, use partnerships as a means for improving their responses to changing business conditions.

Management Skills and Human Factors

Managing an SME in an integrated supply chain is a complex task, and participation in multiple chains adds to the complexity. Rapid changes in the business environment, shorter product life cycles, and increasing customer demands require a robust management team that is willing and able to respond rapidly to changes.

Recommendation. Although extensive formal planning may not be justified, it is becoming imperative that small and medium-sized manufacturing enterprises periodically pause from the rush of daily business to survey the business environment of rapidly changing technologies and customer requirements and develop brief, formal business plans.

Recommendation. Small and medium-sized manufacturing enterprises should (1) assess and strengthen their management capabilities; (2) create a corporate environment conducive to the flexibility, change, evolving skills, and learning required by integrated supply chains; (3) integrate their own supply chains; (4) learn to deal effectively with risk; (5) develop the people skills required to integrate effectively with customer supply chains; and (6) engender a shift in corporate attitudes about supply chains from "what's in it for me" to "how can we maximize the common good." Because all of the requisite skills are rarely resident in a single entrepreneur, SMEs should, whenever possible, increase the breadth and depth of their management teams.

Technology

Technology is playing an increasingly significant role in the success or failure of SMEs. Although up-to-date manufacturing and process technologies are critical, they are no longer the only required technologies. Information technology has become a key to operating success. Internet technologies alone are changing the mechanisms of communication, marketing, selling, buying, and generating revenue.

Recommendation. Small and medium-sized manufacturing enterprises (SMEs) should keep abreast of customer expectations regarding on-line responsiveness and use e-business service providers to assist them in creating and operating low-cost Web sites for displaying products, accepting orders, and answering frequently asked questions. SMEs without an on-line presence may find themselves at a strong competitive disadvantage.

Recommendation. Although the highest levels of communication capabilities can provide incredible competitive power, they are too complex and costly for most small and medium-sized manufacturing enterprises (SMEs). These technologies should be monitored closely, however, because their costs and ease of implementation are improving dramatically. Internet technologies can provide many of these capabilities today at far lower cost, and SMEs should take advantage of these easy-to-use technologies.

Recommendation. Despite significant media coverage of the capabilities of business management systems, small and medium-sized manufacturing enterprises should evaluate, but generally defer, purchasing enterprise resource planning and supply chain integration software until prices come down, these systems are easier to install and use, and the benefits of specific systems have been more thoroughly validated.

Recommendation. Regardless of the level of integration, senior management in small and medium-sized manufacturing enterprises should take the lead in using Internet technologies within their companies. They should closely monitor changes in information technology and invest now in basic capabilities, plan for future investments to support their competitive position, and study how and when to integrate their systems with those of other supply chain participants. Senior management should define data requirements and closely manage the implementation of appropriate data management and electronic communication capabilities.

Recommendation. Despite the increasing importance and glamour of Internet-based technologies, small and medium-sized manufacturing enterprises should not ignore up-to-date manufacturing and process technologies. They remain essential for success.

Recommendation. As supply chain integration requirements and the need for new technologies increase the financial requirements imposed on small and medium-sized manufacturing enterprises, they should integrate their own supply chains to reduce redundant inventories and excess manufacturing capacities, thereby freeing cash for other investments.

CHARACTERISTICS OF SUCCESSFUL SMALL AND MEDIUM-SIZED MANUFACTURING ENTERPRISES

Based on interviews conducted for this study and the experience of committee members, successful SMEs tend to:

- choose customers carefully
- react appropriately to salient events that can define success or failure
- establish strategic alliances and partnerships with customers and suppliers
- cater to customers' needs
- focus on quality
- treat employees as valuable assets
- select and monitor appropriate metrics
- document business and manufacturing processes
- use the Internet for business communications and education
- share information with supply chain partners

ASSISTANCE FOR SMALL AND MEDIUM-SIZED MANUFACTURING ENTERPRISES

Not-for-profit MEC/TRPs (manufacturing extension centers and technical resource providers), chartered specifically to provide advice and counsel to SMEs, can be found in virtually every city and region in the United States. Although these organizations are extremely helpful to SMEs, the committee found that not all of them are fully capable of helping SMEs compete successfully in a rapidly changing integrated supply chain environment, nor are they consistently proficient in providing guidance to SMEs attempting to integrate their own supply chains. Specifically, MEC/TRPs must develop a standard set of supply chain best practices for SMEs and implement uniform integrated supply chain support programs at all of their centers. These programs must be of uniformly high quality because supply chain integration typically involves multiple companies in scattered locations, and inconsistencies among local programs and levels of support can make integration efforts difficult. MEC/TRPs will require sufficient public and private funding so that they can focus their efforts, not on fund raising, but on this important new mission without detracting from other critical SME support operations.

CONCLUSIONS

The specific recommendations in the body of this report are the true conclusions of this study. Nevertheless, some salient conclusions can be drawn from the report as a whole. In the context of converging trends in supply chain integration, technology, and logistics, which are resulting in dramatic increases in low-cost global competition and substantial demands for investment, small and medium-sized manufacturing enterprises must take the following key steps:

- engage in meaningful strategic planning, not just budgeting
- increase their financial, managerial, and technological strengths
- add value to their products and integrate more closely with their customers
- integrate their own supply chains to reduce costs and improve performance

These responses will not, by themselves, ensure competitiveness, but they are essential for the successful participation of small and medium-sized manufacturing enterprises in modern integrated supply chains.

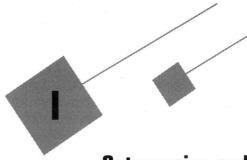

Outsourcing and Supply Chain Integration: Remaking American Industry

"THE CENTRAL EVENT of the twentieth century is the overthrow of matter," wrote George Gilder in his landmark book *Microcosm*. "In technology, economics and the politics of nations, wealth in the form of physical resources is steadily declining in value and significance. The powers of mind are everywhere ascendant over the brute force of things." We live in a world where the most powerful corporation is no longer the one with the biggest factories or the most real estate—but the one with the ability to rapidly turn ideas and thinking into new products, new services and new businesses (*Wall Street Journal*, January 27, 1999).

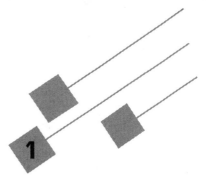

Introduction

The estimated 330,000 U.S. small and medium-sized manufacturing enterprises (SMEs) have a substantial economic impact (Carr, 1998). Defined as having fewer than 500 employees, SMEs are important to the nation because they account for 98 percent of all manufacturing plants, employ two-thirds of the nation's 18 million manufacturing workers, generate more than half of the total value added in the manufacturing sector, and are the source of many innovations in technology (Weber, 1997).

SMEs typically provide capabilities that their larger customers do not have or cannot cost-effectively create internally, such as:

- greater agility in responding to changes in technologies, markets, and trends
- greater efficiency due, in part, to less bureaucracy
- greater initiative and entrepreneurial behavior on the part of employees resulting in higher levels of creativity and energy and a strong desire for success
- access to specialized proprietary technologies, process capabilities, and expertise
- shorter time to market because operations are small and focused
- lower labor costs and less restrictive labor contracts
- spreading the costs of specialized capabilities over larger production volumes by serving multiple customers
- lower cost customized services, including documentation, after-sales support, spare parts, recycling, and disposal

Thus, SMEs can provide a wealth of value if they are used effectively in integrated supply chains.

The term "SME" in this report refers only to those small and medium-sized companies actively involved in manufacturing that serve as suppliers to higher tier suppliers or to original equipment manufacturers (OEMs) (i.e., manufacturers that build products for end users rather than components for use in other products). SMEs vary greatly in terms of size, industry, capabilities, and financial strength. They range, for example, from sophisticated 500-person operations that design and fabricate electronic products to several-person local machine shops.

STUDY OBJECTIVES AND APPROACH

The nature of business competition is changing rapidly and with it the supply chains that support OEMs. In response to competitive pressures, U.S. manufacturers are purchasing increasing amounts of goods and services from outside suppliers and are increasing their efforts to integrate their supply chains to improve performance. A 1998 survey found that 80 percent of manufacturers have formal supply chain management programs or plan to start them in the next year (Manufacturing News, 1998). In this context, the Manufacturing Extension Partnership (MEP) of the National Institute of Standards and Technologies (NIST) and the Robert C. Byrd Institute (RCBI) requested that the National Research Council study the new roles and challenges for SMEs resulting from increasing supply chain integration. The Committee on Supply Chain Integration, formed under the direction of the Board on Manufacturing and Engineering Design, was asked to perform the following tasks:

- Identify and analyze state-of-the-art supply chain integration concepts.
- Define the requirements for successful SME participation in integrated manufacturing supply chains.
- Define the gaps between the requirements and capabilities of SMEs.
- Suggest strategies to assist SMEs in developing the capabilities necessary for successful participation.

Despite the strong overall trend toward increased supply chain integration, the extent of integration varies greatly from industry to industry. Defense industries, for example, are not allowed to make decisions based on the overall good of supply chains until changes have been made in the Federal Acquisition Regulations (FAR) and contract law. Contract law, especially in accordance with FAR, flows down fiduciary responsibility (i.e., each supplier is directly accountable for its portion of the work as defined in the contract or purchase order). Thus, efforts to optimize overall supply chain performance for the mutual benefit of all participants are generally not allowed unless specifically permitted in the contract.

The committee supplemented its expertise and gained a deeper understanding of issues relating to SMEs and supply chain integration in several ways. First, the committee met with a group of representatives (mostly chief executive officers) of successful SMEs, who presented overviews of their companies and discussed their experiences with supply chain integration. The results of these discussions, combined with the collective experience of the committee members, were distilled into the list of characteristics of successful SMEs presented in Chapter 11.

To test their experience base, the committee developed a brief questionnaire asking SMEs about their technological capabilities and relationships with customers. Field agents of MEP and RCBI administered the survey to 99 SMEs. The results of the questionnaire, although not statistically significant, were useful for obtaining a deeper understanding of the issues confronting SMEs and their customers. The results are discussed in Chapter 6. (A copy of the questionnaire and a summary of the data are included in Appendix A.)

Although each SME has its own definition of success, the committee defined the phrase *successful supply chain participation* in a manner similar to the traditional definition of *business success* (i.e., participation in which benefits to the corporation substantially exceed the costs).

Although large suppliers are frequently cited as examples in this report, the lessons and principles generally pertain to SMEs as well. Some of the trends may be slower in coming to SMEs, which are often in the lower tiers of the supply chain. However, SMEs should study these and similar cases carefully so that they can improve their performance both as suppliers and as managers of their own supply chains.

This report is divided into two parts. Part I introduces the concepts of outsourcing and supply chain integration. Part II identifies the requirements imposed on SMEs by integrated supply chains, compares them with the capabilities of SMEs, and recommends courses of action that can enable SMEs to fill their own specific capability gaps.

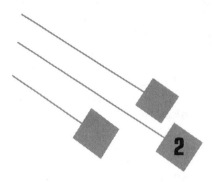

Manufacturing Supply Chains

Manufacturing can be defined as an activity which, utilizing a variety of capabilities, adds value to a material, thereby making possible different uses of that material. Each step in the manufacturing process adds value. The first manufacturers were probably artisans who worked by themselves to design and create products. They served as both supplier and manufacturer, gathering and managing the resources, and applying various processes to add value to the materials. Over time, manufacturing progressed to a series of specialists, each of whom supplied or added specific amounts and types of value. The benefits of this division of labor were that (1) more resources could be brought to bear on the task of adding value and (2) specialization tended to reduce costs and increase the efficiency, consistency, and quality of each operation.

As manufacturing industries developed, their products and processes became more complex. By 1920, well developed relationships between customers and suppliers facilitated the emergence of "mass production." At that time, many large OEMs were vertically integrated, purchasing only commodity products (e.g., steel, glass, and paint) and components that required specialized facilities or technology.

OUTSOURCING

As worldwide competition increased in the 1980s and 1990s, manufacturing profitability came under severe pressure. U.S. manufacturers, having substantially reduced internal costs, began searching for additional opportunities for increasing their competitiveness. This effort led to the so-called "outsourcing movement." As manufacturers analyzed the amount, costs, and types of value added to products in their own

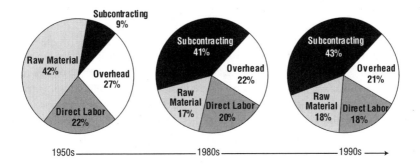

FIGURE 2-1 Increase in subcontracting in the defense industry (percentage of total product cost). Source: The Boeing Company, 1997.

facilities and compared them to capabilities available from outside suppliers, it became apparent that many elements of value could be purchased, or "outsourced," ("subcontracted" in the defense industry) more cost effectively. By the early 1990s, reengineering, downsizing, and outsourcing had become common business practices. Figure 2-1 shows the subcontracting trend in the defense industry.

The trend toward outsourcing was reinforced during the 1990s by Wall Street, as securities analysts increasingly focused on return on assets as a measure for valuing corporations. In response, OEMs have increasingly sold off parts-fabrication operations, which tend to require substantial investments in assets, and have focused on final assembly and services. As a result, OEMs have become increasingly dependent on outsourcing.

An estimated 22 percent of the world's gross domestic product consists of raw materials, components, and subassemblies purchased by OEMs. Thus, reducing the cost of these materials and making the procurement process more efficient can have a substantial effect on the global economy, as well as on the competitiveness of individual OEMs.

Case Study: Chrysler Corporation

Late in the twentieth century, two competitors in the automotive industry, Chrysler Corporation and General Motors (GM), made dramatically different choices regarding their respective supply chain structures. GM, which was consolidated into a massive, vertically integrated corporation early in the century, produced most of its components and subassemblies internally, although its Delphi parts operation was spun off in

1999. GM benefited by having direct control of its parts-making operations and from economies of scale.

Chrysler Corporation (now part of DaimlerChrysler AG), partly in response to its desperate financial condition in the 1980s, began leveraging its capabilities through extensive outsourcing. The company then reduced the number of redundant participants in its vast supply chain, providing more work for the remaining participants in return for lower prices. Remaining participants were also given increased responsibility for quality and just-in-time delivery. With this approach, Chrysler reduced its costs, its in-house inventories, and the number of product defects and increased the efficiency of its internal assembly lines. In an attempt to gain even more benefits from its suppliers, Chrysler is increasingly involving them in product development and mandating annual improvements in production efficiency. This has resulted in further cost reductions and faster development of increasingly innovative products.

Relationships between OEMs and suppliers in the U.S. auto industry have traditionally been adversarial. Products were designed with little input from suppliers; suppliers were selected by competitive bidding based almost solely on price; and purchasing agreements allowed suppliers little flexibility. Although the transition from an adversarial approach to a partnership arrangement with a free flow of ideas has been exceedingly difficult, Chrysler has made substantial progress. Instead of competitively rebidding supply contracts every two years, most of Chrysler's agreements now extend over the life of the model, and sometimes beyond. Essentially, Chrysler's business is the supplier's to keep as long as the supplier performs well on the current model and meets cost targets on the next.

The results of this approach have been dramatic (Dyer, 1996):

- New vehicle development time was reduced from 234 weeks in the 1980s to approximately 160 weeks in the mid-1990s.
- The cost of developing a new vehicle dropped by an estimated 20 to 40 percent.
- Chrysler's average profit per vehicle increased from $250 in the mid-1980s to $2,100 in the mid-1990s.
- Under the old system, 12 to 18 months of the development process were devoted to soliciting bids, analyzing quotes, rebidding, negotiating contracts, and tooling suppliers for production. Additional time was required to solve problems encountered by suppliers who, having bid successfully, attempted to manufacture components they had not designed. Under the new system, suppliers are involved throughout the process, from initial concept through

concurrent design and volume production. This has reduced parts incompatibility and allowed more time for resolving problems despite the shorter overall cycle time.

- With early supplier involvement, prototypes are completed earlier, defects are found faster, and hard tooling is purchased only after most problems have been resolved. Tooling is purchased as much as 12 months closer to the initial production date, reducing the amount of tool rework, as well as the amount of capital invested.

- As a result of longer commitments, suppliers have increased their own investments in assets dedicated to Chrysler, including plants, equipment, systems, processes, and people. Nearly all of them have purchased Chrysler's preferred computer-aided three-dimensional interactive application (CATIA) system, which is designed to enable concurrent engineering. With CATIA, the 1998 Concorde and Intrepid were designed and developed with an almost paperless process, reducing the development time for 1998 models by eight months and saving more than $75 million (Hong, 1998). Many suppliers have also relocated their facilities in closer geographical proximity to Chrysler plants.

The following factors were crucial to the transformation of Chrysler's supply chain (Dyer, 1996):

- Strong, visionary leadership that drove the change to collaborative approaches for jointly creating value.

- Multifunctional teams ("platform teams"), including suppliers' engineers, are now responsible for the product line, from concept through manufacturing, which has shortened the product development cycle. To speed up decision-making, platform teams include representatives of multiple functions, including engineering, manufacturing, finance, marketing, and procurement. This approach has stabilized priorities and reduced the conflicting demands to suppliers that were inherent in the old sequential development process.

- Platform teams select suppliers early in the concept stage from lists of prequalified suppliers with the best track records and the most advanced engineering and manufacturing capabilities. Suppliers are given major responsibilities for component design, cost, quality, and on-time delivery. Suppliers indicate that this approach gives them greater flexibility to develop effective solutions to problems as they arise.

- Changing from competitive bidding to target costing helped to shift the relationship with suppliers from a zero-sum to a positive-sum game and created an atmosphere in which OEMs and suppliers work together to achieve the target. This atmosphere has fostered a growing trust, which is essential for a successful partnership.
- SCORE (the supplier cost reduction effort) continues to motivate suppliers to reduce costs and increase value. The program commits Chrysler to encouraging, reviewing, and acting on supplier ideas quickly and fairly and to sharing the benefits equitably. Suppliers are required to offer annual suggestions equaling 5 percent of their sales to Chrysler. Suppliers can claim up to half of the savings for themselves or share more of them with Chrysler to improve their supplier performance rating and perhaps obtain additional future business. Delighted suppliers throughout the chain have responded with a continuing flow of ideas for improvement.
- To facilitate interactions, Chrysler established a common e-mail system with suppliers and schedules face-to-face meetings with them on a regular basis.

The late 1990s merger that created DaimlerChrysler is an example of the global trend toward consolidation in automobile manufacturing. Potential cost savings were a key motivator for the merger, and the new corporation immediately announced that it intended to optimize supplier performance further, building stronger relationships with key suppliers to more effectively manage the entire chain down to the raw material level. First-tier suppliers are expected to play increased roles in managing their own supply chains, producing better products at lower cost and taking the lead in programs, ranging from research and development to the design and production of complete modules.

Thus, Chrysler has achieved a temporary advantage over its competitors through careful outsourcing and management of an extensive supply chain. However, in so doing, they have also helped to create suppliers with extensive competence and muscle. This new situation raises several questions. Will Chrysler's supplier/partners, such as traditional seat suppliers Johnson Controls, Inc., and Lear Corporation, use their positions to seize a greater percentage of Chrysler's revenues and profits? If recent moves away from vertical integration were right for Chrysler, why has vertical integration by the seat suppliers increased their business success? Under what conditions is outsourcing better than vertical integration?

The answers to these questions appear to depend on complex, evolving, industry-specific business criteria, such as which manufacturer can provide the lowest cost capabilities and which capabilities must be retained internally for competitive advantage. Traditionally, automakers competed on the basis of new products and manufacturing efficiency. Today, they are seeking allies to help them revamp their distribution systems, cut costs and inventories, and gain broad access to the Internet car-consumer market. Thus, one key to success is the ability to adapt the supply chain rapidly to changing conditions.

Benefits of Outsourcing

OEMs may reap the following benefits from outsourcing:

- improved focus, quality, and simplification of remaining in-house operations
- lower cost manufacturing operations, including reduction of in-house inventories
- shorter product realization cycles (faster time to market) and lower product development costs if suppliers are directly involved in product design
- access to capabilities and technologies that could not be readily developed or cost-effectively acquired
- additional manufacturing capacity and faster response to changing market demands, often without additional capital investment by the OEM

Despite these benefits, OEMs may elect to retain aspects of vertical integration, such as capabilities that provide sustainable competitive advantage or identify a product with the OEM.

Outsourcing has also created substantial benefits for suppliers. Some of these opportunities, such as decisions by IBM and others in the 1980s to outsource PC operating systems to Microsoft and microprocessors to Intel, have led to the creation of huge fortunes. For example, automakers traditionally made or bought most of the parts for automotive seats and assembled the seats themselves. However, trends in the 1990s toward cost reduction, outsourcing, and reduction in the number of suppliers created expanded opportunities for surviving suppliers. Johnson Controls and Lear, which responded with cost controls, growth strategies, and acquisitions, have taken over the assembly task and are now the dominant suppliers of fully assembled modular seating systems to GM, Ford, and DaimlerChrysler. The modules are delivered in sequence and installed directly into vehicles as they move along the assembly line.

On-time delivery, low costs, high quality, and low error rates are critical in this competitive, make-to-order business. Leveraging their low costs and global capabilities, Johnson Controls and Lear are now

acquiring makers of other interior components and are positioning themselves to supply fully integrated automotive interiors. For example,
in March 1999, Lear purchased United Technologies Corporation's
automotive-parts business, enabling it to deliver modules that include
instrument panels, electrical wiring systems, and electronic seat controls.
The automakers, in turn, are further integrating with these suppliers by
hiring them to help in the design process, a high-margin activity that
benefits both the OEMs and suppliers.

SUPPLY CHAINS

As manufacturing operations become increasingly specialized and
complex, suppliers are relying more heavily on their own suppliers (often
referred to as "lower tier" suppliers), who, in turn, rely on still others.
This extended, chain-like system of companies, brought together to fulfill
an end-customer demand, has come to be called a "supply chain."
Figure 2-2 shows a typical supply chain structure.

The committee defined a *supply chain* as an association of customers
and suppliers who, working together yet in their own best interests, buy,
convert, distribute, and sell goods and services among themselves resulting in the creation of a specific end product. Thus, every company is
necessarily part of at least one supply chain. It is not a matter of choice.
The chains in which a company participates are defined by (1) the OEM
and its suppliers, who together provide all of the capabilities required
to create and support the end product, and (2) the customers who purchase it.

The supply chain includes all of the capabilities and functions required to design, fabricate, distribute, sell, support, use, and recycle or
dispose of a product, as well as the associated information that flows up
and down the chain. Supply chains are typically comprised of geographically dispersed facilities and capabilities, including sources of raw materials, product design and engineering organizations, manufacturing plants,
distribution centers, retail outlets, and customers, as well as the transportation and communications links between them. OEMs typically have a
supply chain for each product line, although the same capabilities are
often used in multiple chains. Many suppliers participate in the supply
chains of more than one OEM.

Many companies are adopting a supply chain structure similar to that
of the construction industry, in which general contractors subcontract
most of the work on an *ad hoc* basis. Contractors solicit bids for the skills
and capabilities required for a specific job from subcontractors they know
and trust. Subcontracts are typically awarded to those with the best combination of capabilities and prices, not necessarily just the lowest prices.

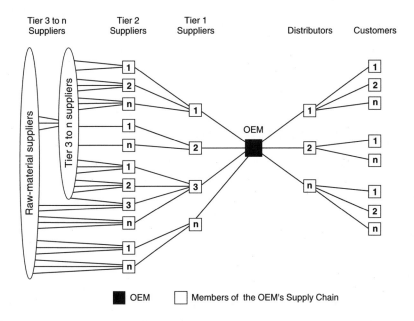

FIGURE 2-2 Structure of a typical supply chain. Source: Adapted from Lambert et al., 1998.

The contracts/partnerships that create this "virtual corporation" last for the duration of the job. New, *ad hoc* teams are formed and disbanded, as needed, for each new job or product. Similar examples can be found in the defense industry, where prime contractors create *ad hoc* teams, bringing together only the skills required to win and execute a specific contract. Subcontractors, in turn, have their own suppliers, who are also part of the chain.

The *ad hoc* supply chain structure has limited value. It tends to work best in industries in which (1) jobs or business opportunities are episodic and somewhat unpredictable, rather than continuous, (2) the required capability or skill mix varies from job to job, and (3) the costs of retaining a full spectrum of skills cannot be justified. For most industries, however, building long-term relationships based on trust and a high level of integration yields greater benefits. Developing trust within a supply chain takes time and effort. Even in the defense and construction industries, benefits can often be maximized through the nurturing of long-term relationships (within the letter of the law), even if the skill set is only used on a contract-by-contract basis.

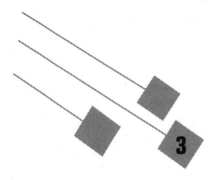

Supply Chain Integration

Most OEMs no longer compete solely as autonomous corporations. They also compete as participants in integrated supply chains. This revolution, which is changing the ways products are designed, produced, and delivered, has the potential to alter the manufacturing landscape as dramatically as the industrial revolution or the advent of mass production. This chapter describes the changing nature of supply chains and efforts to optimize their performance.

In the past, OEMs typically drove down the cost of purchased materials through aggressive negotiations, imposing terms and conditions that minimized supplier profitability and often left suppliers in a weakened condition. More recently, OEMs have begun to adopt a strategic partnership approach, which recognizes that increased, sustainable benefits can accrue from long-term relationships between participants in the supply chain (a win-win situation). This approach considers total life-cycle costs over multiple iterations of a product, with the goal of increasing mutual benefits for all participants in the long run.

SUPPLY CHAIN MANAGEMENT

In this era of competition among supply chains, the success of a corporation is increasingly dependent on management's ability to integrate the company's networks of business relationships. *Supply chain management* (see Figure 3-1) has been defined as the integration of key business processes, from raw-material suppliers through end users, that

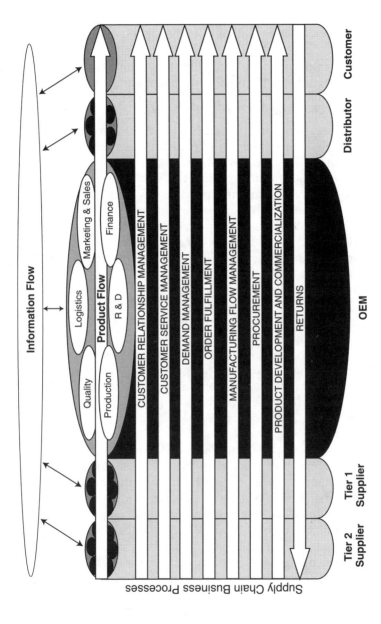

FIGURE 3-1 Supply chain management: integrating and managing business processes among participants throughout the supply chain. Source: Adapted from Lambert et al., 1998.

provide products, services, and information that add value for customers and other stakeholders (Lambert et al., 1998).

Supply chain management makes use of a growing body of tools, techniques, and skills for coordinating and optimizing key processes, functions, and relationships, both within the OEM and among its suppliers and customers, to enable and capture opportunities for synergy. An OEM's competitive advantage is highly dependent on this integrated management function. Supply chain management attempts to combine the best of both worlds, the scale and coordination of large companies with the low costs, flexibility, and creativity of small companies.

The focus of supply chain management must evolve in response to changing business environments and evolving product life cycles. Different interactions among participants are required during each phase of the product life cycle, from inception through recycling. The supply chains for products in new markets must be flexible to respond to wide fluctuations in demand (both in quantity and product mix). Products in mature, stable markets require supply chains that can reliably deliver products at low cost. Thus, effective supply chain management must be responsive to these changing conditions to ensure that the supply chain evolves accordingly.

For example, marketing excellence used to be the primary source of Procter & Gamble's (P&G's) dominance of the consumer products industry. However, as P&G expanded its product and service offerings in response to market opportunities, the increased complexity of these offerings created difficulties in meeting the needs of retail partners and customers. Traditional marketing strategies involving in-store sales and price promotions created great variations in product demand. To meet heavy short-term marketing-induced peaks in demand, P&G invested in huge manufacturing capacities, inventories, warehouses, and logistics capabilities.

In response to these problems, P&G modified its supply chain focus and remade itself through a series of innovative initiatives. Working both internally and with suppliers and customers, the company created a heralded partnership with Wal-Mart, virtually eliminated price promotions, and streamlined its logistics and continuous replenishment programs. These initiatives reduced variations and uncertainties in demand, thereby reducing the need for surge production capacities and large inventories. Thus, by evolving their primary supply chain focus from marketing to production, inventories, and logistics in response to changing business requirements, P&G was able to reduce costs, meet customer demand, and build strong, coordinated relationships with retail partners and customers.

CONCEPT OF INTEGRATION

An *integrated supply chain* can be defined as an association of customers and suppliers who, using management techniques, work together to optimize their collective performance in the creation, distribution, and support of an end product. It may be helpful to think of the participants as the divisions of a large, vertically integrated corporation, although the independent companies in the chain are bound together only by trust, shared objectives, and contracts entered into on a voluntary basis. Unlike captive suppliers (divisions of a large corporation that typically serve primarily the parent corporation), independent suppliers are often faced with the conflicting demands of multiple customers. (The technological and investment problems faced by SMEs in attempting to deal with these conflicting demands are discussed in Chapter 9.)

All supply chains are integrated to some extent. One objective of increasing integration is focusing and coordinating the relevant resources of each participant on the needs of the supply chain to optimize the overall performance of the chain. The integration process requires the disciplined application of management skills, processes, and technologies to couple key functions and capabilities of the chain and take advantage of the available business opportunities. Goals typically include higher profits and reduced risks for all participants.

Traditional unmanaged (or minimally managed) supply chains are characterized by (1) adversarial relationships between customers and suppliers, including win-lose negotiations; (2) little regard for sharing benefits and risks; (3) short-term focus, with little concern for mutual long-term success; (4) primary emphasis on cost and delivery, with little concern for added value; (5) limited communications; and (6) little interaction between the OEM and suppliers more than one or two tiers away. Integrated supply chains tend to recognize that all parties should benefit from the relationship on a sustainable, long-term basis and are characterized by partnerships with extensive and open communications. A well integrated system of independent participants can be visualized as a flock of redwing black birds flying over a marsh. Without any apparent signal, every bird in the flock climbs, dives, or turns at virtually the same instant. That is an integrated system! Supply chain members, in a similar manner, must react coherently to changes in the business environment to remain competitive.

Supply chain integration is a continuous process that can be optimized only when OEMs, customers, and suppliers work together to improve their relationships and when all participants are aware of key activities at all levels in the chain. First-tier suppliers can play a key role in promoting integration by guiding and assisting lower tier suppliers. In an example of multi-tier integration, Wal-Mart thoroughly integrated P&G's Pampers product line into its supply chain. P&G, in turn, worked with 3M to integrate its production of adhesive strips with Pampers manufacturing facilities.

Forces Driving Increased Integration

The following worldwide trends and forces are driving supply chains toward increased integration:

- *Increased cost competitiveness.* Having substantially improved the efficiencies of internal operations, OEMs are seeking further cost reductions by improving efficiency and synergy within their supply chains.
- *Shorter product life cycles.* The Model-T Ford, for example, was competitive for many years. A personal computer (PC) is state of the art for less than a year, and the trend toward shorter product life cycles continues.
- *Faster product development cycles.* Companies must reduce the development cycle times of their products to remain competitive. Early introduction of a new product is often rewarded with a large market share and sufficient unit volumes to drive costs down rapidly.
- *Globalization and customization of product offerings.* Customers the world over can increasingly afford and are demanding a greater variety of products that address their specific needs. Mass customization has become the new marketing mantra.
- *Higher overall quality.* Increasing customer affluence and tougher competition to supply their needs have led to demands for higher overall quality.

These increased demands on OEMs for improvements in product design, manufacturing, cost, distribution, and support are being imposed, in turn, on their supply chains.

Case Study: Dell Computer and Fujitsu America

Dell Computer Corporation's success in the past few years and its growth relative to the rest of the PC industry made daily headlines throughout the 1990s. Based on the premise that bypassing resellers, building products to order, and reducing inventories would result in a lower cost, more responsive business, Dell has grown into one of the largest forces in the industry. Nevertheless, it is squeezed into such a narrow business niche that, from some perspectives, its very survival seems tenuous. Dell competes with many capable and, in some cases, lower cost competitors, has virtually no proprietary technology, and must deal with exceedingly robust suppliers, including Intel and Microsoft.

The heart of Dell's success is its integrated supply chain, which has enabled rapid product design, fabrication, and assembly, as well as direct shipment to customers. Inventories have been dramatically reduced through extensive sharing of information, a prudent choice given the risk of technological obsolescence and reductions in the cost of materials that can exceed 50 percent a month. Even with reduced inventories, Dell's strategic use of information has made possible a dramatic reduction in the elapsed time from order to delivery, giving Dell a significant competitive advantage.

Component inventories are monitored weekly throughout the supply chain and, when there are deviations from plan, the sales force steers customers, by means of discounts, if necessary, toward configurations for which there are adequate supplies. Thus, abundant, timely information is used to work the front and back ends of the supply chain simultaneously.

Speed is a critical factor in the computer industry, especially in the area of inventory. In the late 1980s, Dell measured component inventories in weeks. In 1998, they were measured in days. They may soon be further reduced through real-time deliveries so that, as components are used, they are automatically and immediately replaced. The reduction in inventory not only lowers requirements for capital, it also enables rapid changeovers to new product configurations because no old parts must be used up. Faster time to market for new products translates into increased revenues and profits. The change in emphasis from inventory levels to inventory velocity throughout the supply chain has been made possible, in part, by the Internet.

In Dell's new virtual corporation, inventories are reduced by use of timely information; emphasis on physical assets is being replaced by emphasis on intellectual capabilities; and proprietary business knowledge is being increasingly shared in open, collaborative relationships. This extensive integration of the supply chain can be viewed as a shift from vertical corporate integration to a virtually integrated corporation (Magretta, 1998). Vertical integration was essential in the early years of computer manufacturing when the supplier base was not well established and assemblers had little choice but to design and build components and assemble the entire end product in house. Proprietary component technologies were a main source of competitive advantage, although in some cases they had little to do with creating value for the customer. As the industry matured, multitudes of component suppliers became eager to invest and compete in terms of price and innovation. Leveraging invest-ments by these suppliers has freed Dell to focus on delivering complete solutions to its customers. However, because these components are avail-able to all PC assemblers, it has become harder to compete in terms of end-product differentiation. Thus, a high premium has been placed on speed and process efficiency, blurring the traditional boundaries between supplier, manufacturer, and customer. For instance, peripherals, such as monitors, keyboards, speakers, and mice, need not be gathered in one location prior to shipment to the customer. Manufactured by separate suppliers and labeled with the Dell logo, shippers gather them from all over North America, match them overnight (merge-in-transit), and de-liver them as complete hardware sets to customers as if they had come from the same location.

Dell's virtual integration has the following characteristics:

- use of rapid, seamless communication to build direct relationships between customers, OEM, and suppliers
- a clear definition of what Dell does best (i.e., core competencies, including branding, marketing, and selling through direct chan-nels), with partnerships for the rest (capital-intensive and labor-intensive component fabrication processes and services). This en-ables Dell to be highly selective in its capital investments and to focus on activities that create the most value for customers and shareholders
- selection of partners who are best in their respective fields, inviting them to become intimate parts of the business, and holding them to the same exacting quality and performance standards as in-house segments of the business
- use of a minimum number of suppliers, to whom Dell is highly loyal as long as they maintain their leadership in technology, qual-ity, cost, and delivery
- use of the Internet, not just as an add-on to the business, but as an integral part of a strategy to eliminate boundaries between compa-nies and promote effective integration
- less emphasis on guarding intellectual assets and more emphasis on using assets rapidly before they become technologically obsolete

By using a highly integrated supply chain, Dell has enjoyed many of the advantages of vertical integration while simultaneously benefiting from the investments, innovation, efficiencies, and specialization of highly focused suppliers. Although the Dell model does not fit every situation, the lessons of Dell's experience can be extracted and adapted to many other supply chain situations, even for SMEs.

By 1998, the success of the Dell model, as might be expected, was causing problems for competitors, including Fujitsu America, which had large inventories and high shipping costs (*Washington Post*, May 2, 1999). Customers had to wait 10 days for laptops, while competitors were delivering in five. In response, Fujitsu moved its distribution center from Portland, Oregon, to Memphis, Tennessee, and turned distribution over to FedEx Corporation, the parent company of Federal Express. In direct response to orders, FedEx coordinates the shipment of components from worldwide suppliers, oversees the assembly of PCs, and ships them out, all in three or four days. By early 1999, the cycle time on the ground was eight to twelve hours, and the goal was to reduce it to four hours. Fujitsu has essentially eliminated geographic proximity as an issue and has made maximum use of the benefits of globalization, including low cost. Even with the premium price of express shipping, this modification of the Fujitsu supply chain saved the company millions of dollars, slashed inventories by about 90 percent, and increased profits by 25 percent. Most important, these changes have enabled Fujitsu to compete effectively with Dell for Internet sales directly to consumers. However, as is evident from these examples, these innovations in supply chain integration can also impose large burdens on suppliers in terms of responsiveness, inventories, and management of their own supply chains.

COSTS OF INTEGRATION

The costs, complexities, and risks of fully integrating and managing a highly integrated supply chain can be as substantial as the costs of integrating and operating a corporation of comparable size. Thus, most supply chain integration efforts to date have been very limited in scope. Some of the major costs are listed below:

- time devoted to managing, training, and support
- effort devoted to becoming a better customer
- investment in supply chain integration software and compatible information systems throughout the chain
- opportunity costs (i.e., investments in supply chain integration may necessitate foregoing other business opportunities)
- risks of production stoppages

Because the extent of interconnectedness and interdependency makes highly integrated chains increasingly vulnerable to disruptions, the risk of production stoppages should not be overlooked. A highly integrated, interdependent supply chain that consists primarily of sole-source suppliers practicing just-in-time manufacturing with minimal inventories is highly reliant on the timely delivery of quality components and services. Failure by one participant to deliver can rapidly bring other parts of the chain to a halt. This happens, on occasion, even to the best suppliers and logistics providers.

Automakers, for example, who are under constant pressure to reduce costs, have tightened their supply chains to the point that they typically have less than a one-day supply of parts at final assembly facilities. Thus, a breakdown anywhere in the supply chain has the potential of bringing production to a halt (e.g., strikes at two GM parts plants in 1998 resulted in the shutdown of virtually all assembly operations within days, and flooding in 1999 at a single supplier in North Carolina reduced operations at seven DaimlerChrysler and three GM assembly plants to half-shifts due to shortages of a single part).

Potential threats, including storms, power outages, terrorism, computer hackers, disruptions in communications, and equipment breakdowns, can be very difficult to predict and costly to prepare for. In another example, the earthquake that shook Taiwan in September 1999 showed how a power supply disruption in one country can have worldwide reverberations through an entire industry. Damage to two electric power substations was the primary cause of a shutdown of Taiwan's computer-chip industry, which resulted in shortages of components and higher costs in the supply chains of OEMs around the world.

Supply chain participants must individually and collectively assess the probability of production-stopping events and their tolerance for risk, which must be balanced against the savings from increased sole-sourcing, tighter integration with remaining suppliers, and reduced inventories and production capacities. Thus, although good communications and resource sharing can be helpful in preparing for and responding to disruptions, supply chain participants must be careful to avoid unacceptable levels of risk in their zeal for integration.

Recommendation. Small and medium-sized manufacturing enterprises should develop operating strategies based on an appropriate balance between supply chain performance and risk; assess the probability and effects of potential threats to their supply chains; and maintain sufficient (though sometimes expensive) slack, redundancy, and flexibility to keep the potential threats at manageable levels.

BENEFITS OF INTEGRATION

The most sought-after benefit, or return on investment, in supply chain integration is the cost savings that result from reductions in inventory. Inventories can be reduced by increasing the speed at which materials move through the supply chain and by reducing safety stocks. For example, if the costs of maintaining inventory are approximately 1 percent per month and if an integrated supply chain can reduce inventory levels by 30 percent, the savings, shared among the participants, can be substantial.

Another common benefit of supply chain integration is a reduction in transaction costs. If information sharing can reduce the number of transactions and if electronic systems can reduce the cost of each transaction from the $150 cost of a traditional transaction, each participant can realize substantial savings (LaLonde, 1997).

Reductions in supplier redundancy can reduce product costs by increasing production levels at remaining suppliers and reducing the costs of managing the supply chain. Although this can also increase investment and management burdens on suppliers, the delegation of responsibility and authority to entities closer to the action can result in improved decision making, as long as good communications are maintained throughout the chain.

Other potential benefits of supply chain integration are listed below:

- reduced friction, fewer barriers, and less waste of resources on procedures that do not add value
- increased functional and procedural synergy between participants
- faster response to changing market demands
- lower cost manufacturing operations
- lower capital investment in excess manufacturing capacity
- shorter product realization cycles and lower product development costs
- increased competitiveness and profitability

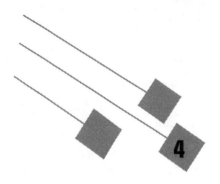

Integration Process

Integrating a supply chain is an incremental process, with priority typically given to the highest potential returns on investment. Based on strategies, needs, and potential returns, different priorities and approaches may be assigned to the supply chains of different segments of a business. The integration process can be expensive and is, in many respects, an exercise in resource allocation.

Many companies adopt an approach that begins at home and gradually works outward through the supply chain. The first step is to make in-house improvements, such as inventory reductions that can reduce working capital, warehousing, and transportation costs. An analysis of in-bound logistics can often reveal opportunities for savings. From there, the integration effort expands outward.

This chapter begins with the topic of supplier selection, introduces approaches to integration, discusses management of the integration process, and identifies management tools that are becoming available to support this effort. The factors critical to success are then identified, and metrics for evaluating progress and performance are suggested.

SUPPLIER SELECTION AND DEVELOPMENT

Supplier selection, development, and integration is a strategic initiative that is undertaken as a part of a company's overall competitive strategy. This strategic approach to outsourcing combines internal core competencies with externally available capabilities and technologies in an attempt to maximize overall corporate and supply chain competitiveness. To achieve these objectives a company must first determine its current and future capability, technology, and capacity needs, map them against its current capabilities, and then assess whether the resulting gaps can best be filled through internal development, acquisitions, or outside suppliers.

If the decision is made to use outside suppliers, the next step involves a worldwide search for competitive suppliers based on the identified capability needs. Company needs should be mapped against the capabilities of potential suppliers. Performance metrics should be established at this stage as a means of assessing candidates and tracking future supplier performance. Suppliers should be carefully selected because the company's commitment, in many cases, will be to a long-term, intimate business relationship. Although some suppliers are selected over other qualified suppliers based on the need to fill government-mandated quotas, most are selected based on combinations of the following factors:

- a track record of demonstrated cost competitiveness and on-time delivery
- possession of proprietary capabilities
- demonstrated management capabilities
- customer support and logistics capabilities
- in-depth quality performance
- willingness to develop jointly seamless processes and eliminate non-value-added activities at the interfaces between customer and supplier
- compatible corporate cultures
- demonstrated financial strength and profitability
- competitive technology and process capabilities
- demonstrated compliance with government regulations
- senior management interest in achieving a sustainable competitive advantage for the supply chain
- willingness to share benefits achieved through supply chain integration

After the pool of potential suppliers has been reduced by a preliminary assessment, the remaining candidates should be subjected to in-depth, on-site risk assessments conducted by a cross-functional team to identify strengths, weaknesses, and deficiencies. Following the final selection, a joint program should be initiated to solve supplier problems, eliminate deficiencies, and establish an open relationship that includes timely feedback and information sharing. This program should include ongoing, systematic supplier development and integration, including joint projects, training, inventory coordination, incentives, and penalties.

Links should be established with second-tier and third-tier suppliers; their key capabilities should be mapped; and their deficiencies should be identified and addressed. Investments in targeted supplier development and integration can result in substantial reductions in product development and order cycle times, as well as improvements in quality and on-time delivery.

Substantial benefits can be realized by including suppliers in the product design process, which may involve new and uncomfortable relationships for many companies. Overcoming these problems will require strong executive support and employee training. Strong, active supplier roles and open sharing of information on the development team will be essential for achieving project goals. Concerns about protecting proprietary information can be addressed by formal confidentiality or nondisclosure agreements.

With increased reliance on sole-source suppliers and expanded levels of supply chain integration, the pressure on each supply chain participant to consistently meet its commitments increases. Replacing unreliable members of highly integrated supply chains and rebuilding required levels of trust and knowledge can be an expensive and painful process.

INTEGRATION BY FUNCTION

Many companies approach integration on a function-by-function basis, focusing first on functions for which integration offers the highest returns. Although the focus differs from industry to industry, inventories, procurement, inbound logistics, manufacturing operations, and distribution of products and services are the functions most frequently integrated. All-inclusive approaches encompass functions ranging from raw materials extraction through manufacturing and distribution to the customer and back. A "closed loop" approach includes asset stripping and the rework or recycling of products returned by customers.

A well integrated supply chain must be open to "functional shiftability" (i.e., the assignment of functional responsibility to members of the supply chain best positioned to perform those functions at the lowest overall cost or in the shortest cycle time). Realignment of such activities within the supply chain should be reflected in a commensurate shift in benefits and risks.

INTEGRATION BY PROCESS

The effort required to identify key functional activities and their interrelationships has caused many companies to change from integrating and managing supply chains by functions to integrating and managing them by process (process management). These companies typically use a business process architecture to analyze processes and supply chain relationships in successive levels of detail. Viewing the supply chain as a set of integrated process capabilities rather than as separate corporations and functions can provide critical insights that can be used to improve performance. In this way, complex activities can be coordinated to great advantage between functions and redundant or non-value-added activities, such as administrative or multiple entries, can be eliminated.

Integration is most beneficial when it occurs across multiple processes that have significant effects on supply chain performance, such as information technology, marketing, and finance. Integration across multiple processes can enable customization of the supply chain according to delivery channels, manufacturing requirements, or market segments.

CRITICAL SUCCESS FACTORS

The following factors are critical for successful supply chain integration (Agility Reports, 1997):

- organizational buy-in, including full commitment by management
- a clear understanding and articulation of identifiable benefits for all parties
- adaptability and openness to changes in work design and organizational structure, consistent with agreed-upon levels of process integration
- effective use of appropriate technologies for communications, data exchange, and product development
- compatibility with the strategic vision of the enterprise

One of the most critical factors is organizational buy-in. Employee responses to integration efforts often range from indifference to antagonism Managers may attempt to "protect their turf," and organizational in-fighting is not uncommon. To some extent these responses are predictable aspects of human nature.

In anticipation of resistance to organizational change, supply chain participants should plan the integration process carefully. First, baseline relationships and processes should be mapped out in detail, an important, but time-consuming process. Second, the system of rewards and sanctions should be modified so that it is congruent with the proposed changes and consistent for all participants, both inside and outside of the corporation. Third, integration should begin on a small scale, using a cross-functional team under the leadership of a process champion, with participants from both customers and suppliers. It can be helpful to separate the team from everyday operations to increase the chances of early success and minimize disruption of nonparticipants until the new approaches have been thoroughly validated. Customer and supplier personnel should be co-located at each other's sites, if possible, to facilitate process integration and communication.

METRICS

Participants in a supply chain are unlikely to achieve their collective goals unless their performance measures (metrics) and incentives are aligned. Hence, metrics and incentives must be clearly and carefully defined, mutually agreed upon, and monitored by all participants. Participants should be held accountable for some of each other's performance measures.

To date, very few companies have succeeded in assessing the performance of their supply chains as a whole. Nevertheless, performance should be measured both on a highly aggregated basis and within specific segments. Although specific metrics must be tailored to the circumstances, the following metrics can be used for high-level assessments:

- profitability
- total sales
- decision response time (the time required to make and implement key decisions throughout the chain)
- return on investment
- return on assets
- technology (the status of and ability to deploy value-enhancing technologies)
- product development time (the elapsed time from concept through initial delivery)
- shared risk (the extent of risk minimization and sharing throughout the chain)
- market share
- planning (the extent to which both strategic and short-range planning are performed in a coordinated and cooperative manner throughout the chain)
- quality (effective planning for and delivery of quality products, including appropriate measurement of results by all participants)
- customer satisfaction
- waste (reductions in scrap, rework, waste, and pollutants from the supply chain and plans for further reductions and recycling)
- transparency (the extent that participants are aware of activities throughout the supply chain)

Detail metrics should be selected for monitoring key functions, processes, and capabilities throughout the chain. Each participant's detail metrics should roll up to an overall measure of supply chain performance. The roll-up should be designed to show how each participant in the supply chain contributes to overall performance. Otherwise, individual firms may take unilateral action to improve their financial or competitive position, which may compromise the performance of the supply chain as a whole. Although detail metrics differ from industry to industry, the ones listed below, which are often used by large corporations to monitor their internal operations, should be considered.

General:
- time to market (the speed at which the organization identifies and delivers new products to the marketplace)
- inventory levels and capacity utilization
- market to collection (the speed with which receivables are converted to cash)
- customer services (measures of after-market support to customers)
- management for results (the efficiency and effectiveness of management)
- infrastructure (the up-time and efficacy of information systems, training, and other support processes)
- return to available (measures of velocity, asset utilization, costs and revenue generated from reworking the product in the returns channel)

Delivery:
- delivery-to-commitment date (measures of meeting commitments)
- lead time (customer waiting period)
- faultless installations (installation errors; customer call-backs)
- faultless invoices (error rates in processing invoices)
- forecast accuracy (accuracy in predicting market demand)
- customer inquiry resolution time (elapsed time to resolve customer inquiries)

Flexibility and responsiveness:
- response time (measures of the responsiveness of supply chain processes or functions)
- productivity flexibility (measures of productivity changes as a function of market demand)
- replanning cycle (time required to create and implement modified production plans)

- release-to-ship date (elapsed time between release of new or modified products and their shipment to customers)
- materials lead time (elapsed time between placement of an order and receipt of materials)

Logistics:
- logistics cost (product distribution costs)
- obsolescence (reduction in inventory value due to expiration of shelf life or changes in technology)
- warranty costs (rework, repair, replacement, shipment, and legal costs associated with defective products delivered to customers)

Asset management:
- cash-to-cash cycle (elapsed time between payment for goods and services used to produce the product and receipt of payment from the customer)
- inventory days of supply (inventory levels divided by average quantity used or shipped per day)
- inventory aging (measures of elapsed time that perishable or potentially obsolete goods remain in inventory)
- days of sales outstanding (accounts receivable divided by average daily sales)
- asset turns (frequency of inventory replenishment per year)
- ship-to-invoice cycle (delay in billing customer after product shipment)

Standards developed by associations and governing bodies provide a common language for communications, which can be used to increase awareness of supply chain performance. Standard techniques, such as statistical process control (SPC), can be useful for monitoring individual processes. However, although certification and conformance to standards is important, they are not always sufficient for measuring some aspects of supply chain performance.

Benchmarking is another technique that can be used for measuring, understanding, and communicating the dynamics of supply chain performance. A detailed discussion of benchmarking in integrated supply chains may be found in Chapter 13 of Robert Camp's book, *Business Process Benchmarking* (1995).

Recommendation. Small and medium-sized manufacturing enterprises (SMEs), as participants in supply chains and as integrators of their own supply chains, should study the process of supply chain integration. Based on a thorough analysis, each participant should then develop an internal business case for participating in particular supply chains and decide, in conformance with formal supply chain agreements and the needs of the chain, the extent to which they will integrate with their customers and suppliers.

Recommendation. Small and medium-sized manufacturing enterprises should use benchmarks and metrics to monitor supply chain performance. Metrics and incentives or penalties should be carefully aligned to encourage optimal supply chain performance. However, because collecting, reporting, and analyzing data is expensive, supply chain participants should select only the metrics required to meet their needs, monitor and analyze the data carefully, and, most importantly, follow through with appropriate action to improve performance.

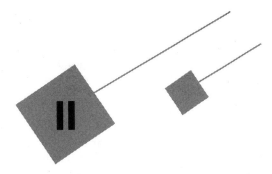

Small and Medium-Sized Manufacturing Enterprises in Integrated Supply Chains

The supply chain in 2020 dominates the manufacturing landscape, with hardware and components manufactured by a whole new tier of job shops and specialty houses—virtual corporations made up of groups of knowledge workers (Owen, 1999).

SMEs cannot ignore the supply chain revolution and remain competitive. Although the outsourcing trend is providing increased opportunities for suppliers, trends toward globalization and increased supply chain integration pose serious challenges. Part II of this report discusses these trends, describes their effects on SMEs, and suggests strategies for achieving competitive advantage.

Each SME selects or creates the supply chains in which it participates. A comparison of the differences between the requirements for participation in OEM supply chains and the unique capabilities of an SME and its own supply chains often reveals deficiencies or gaps the SME must address. In Chapter 5, the committee identifies some of the capabilities of SMEs, based on survey data. Chapters 6 through 10 address specific requirements, constraints, and capability gaps that SMEs must address to succeed. Competitive cost, quality, service, and delivery are the traditional fundamental capabilities required of any supplier, but successful participation in today's integrated supply chains requires more. Technology and management skills, for example, are becoming critical. The requirements and capabilities identified in these chapters are demanded in various combinations and on various time scales depending on the industry and the specific needs of the supply chain.

Some SMEs have been thriving in this new environment. After interviewing executives from a number of successful SMEs, the committee identified the characteristics that make them successful (Chapter 11). Chapter 12 introduces a variety of resources to help SMEs overcome their constraints and fill their capability gaps.

Competing demands can create difficult investment choices for SMEs that participate in multiple chains. Selecting the right technologies can be critical for success. The associated costs of new technologies come early, have a 100 percent probability of occurring, and are generally easy to measure. The benefits, including savings and increased competitiveness, however, come later, may not be fully realized, and are sometimes difficult to measure. Nevertheless, an unwillingness to invest, take risks, and reposition the enterprise in response to the evolving business environment may lead to business failure.

Faced with the simultaneous challenges of buying from large suppliers with concentrated market power and selling to equally large and powerful OEMs or prime defense contractors, SMEs may feel as though they are being squeezed in a vise. The challenge to the small manufacturer is to find ways to participate successfully under these conditions. Because of the tremendous diversity of integrated supply chain requirements and SME capabilities, there can be no universal formula for SME success. However, the recommendations presented in the following chapters, properly applied, can improve the odds for SME competitiveness and profitability.

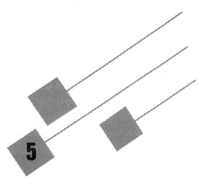

Capabilities of Small and Medium-Sized Manufacturing Enterprises

This chapter presents the results of three surveys, one conducted by the committee, the other two conducted by the Georgia Manufacturing Extension Alliance, to identify trends in the evolving capabilities of SMEs. Although the sizes, industries, and supply chain requirements of the participants vary greatly, the results reveal gaps that must be addressed by SMEs if they are to remain competitive.

COMMITTEE SURVEY

In the summer of 1998, the committee administered a questionnaire to 99 SMEs (1) to gain a better understanding of the current practices and capabilities of SMEs involved in supply chain integration and (2) to identify potential shortfalls in these capabilities. The questionnaire and a summary of the data can be found in Appendix A. The first level of analysis was conducted on the total sample. Subsequent analyses focused on responses as a function of annual revenues and the concentration of the customer base.

First-Level Analysis and Observations

The use of electronic business transactions between customers and SMEs was very limited. Only 11 percent of customers placed orders electronically, but the data was greatly skewed by a few "large" SMEs with electronic capabilities. Thus, although electronic data interfaces and Internet ordering are increasingly cited as important for supply chain integration and although large corporations are investing extensively in electronic capabilities, based on this mid-1998 sample, there was still little use of electronic business among SMEs. The extent to which SMEs are dealing with small customers who are less likely to have this capability is not known. The results of the survey may indicate that SMEs need help in terms of investments, technical know-how, and understanding of the value of electronic business transactions.

SMEs in the survey were generally better equipped in terms of computer-aided design (CAD); 74 percent had CAD capabilities. Approximately half reported using SPC, computer-aided manufacturing (CAM), materials and resource planning (MRP), and hazardous material (HAZMAT) handling capabilities. Although only 41 percent were certified by the International Standards Organization (ISO), SMEs clearly felt that ISO certification was important; 35 percent of the respondents reported plans to obtain certification.

SMEs characterized relationships with their customers as somewhat more "partner-like" than "adversarial," although it appears from the data that there is substantial room for improvement in this area. The SMEs considered early involvement in product development, receiving production forecasts from customers, and sharing performance data to be very important. They reported less of a need for improvement in payment terms, supplier recognition programs, sharing of cost data, and financing. The implication was that SMEs worry less about operations over which they have control and more about factors that are dependent on customers. Therefore, customers should work more closely with SME suppliers in the areas of product development, information sharing, and performance data feedback.

Analysis Based on Size and Customer Concentration

The median annual sales of survey participants was $7.7 million. The committee divided the sample into two equal parts: SMEs with annual sales above the median ("large") and SMEs with sales below the median ("small"). The committee also assessed differences in survey results between SMEs with a higher concentration of major customers (i.e., more than 34 percent of sales from their top three customers) and companies with a more dispersed customer base (i.e., less than 34 percent of sales from their top customers).

The survey indicated little difference between large and small SMEs in the percentage of customers ordering electronically, indicating, perhaps, that customers were still in the process of adopting these capabilities. Large SMEs considered customer sharing of future product and technology plans slightly more important than small SMEs. Large SMEs

TABLE 5-1 Use of Manufacturing Technologies and Techniques, 1996 and 1994

Technologies and Techniques	1996		1994	
	Current Use	Planned Use	Current Use	Planned Use
PCs, non-manufacturing	96.1%	1.2%	90.6%	3.6%
Material requirements planning (MRP II)	58.5%	27.1%	56.1%	28.2%
Just-in-time manufacturing (JIT)	58.2%	14.5%	61.7%	14.2%
Preventive maintenance	53.6%	28.3%	59.2%	28.3%
Local area networks (LANs)	51.0%	13.9%	31.7%	18.3%
Employee teams	49.3%	25.1%	52.8%	26.6%
PCs, shop floor	45.0%	23.2%	36.6%	25.7%
CAD with computer-aided engineering (CAE)	43.8%	11.2%	39.4%	15.7%
Internet	38.2%	30.5%	n/a	n/a
SPC and statistical quality control (SQC)	37.5%	20.5%	36.2%	23.9%
Electronic business transactions	37.0%	32.0%	35.2%	29.8%
Numerically controlled (NC) or computer numerically controlled (CNC) machines	28.8%	8.6%	28.6%	7.6%
Data collection devices	27.2%	32.5%	23.1%	31.6%
CAD with CAM	23.0%	10.7%	17.0%	13.2%
Manufacturing cells	19.2%	10.2%	n/a	n/a
Computer integrated manufacturing (CIM)	16.1%	17.1%	10.2%	20.3%
Automated material handling	15.7%	18.8%	17.0%	20.2%
ISO 9000/QS 9000 certification	13.9%	28.9%	4.2%	36.6%
Automated in-process inspection	11.1%	16.6%	8.9%	15.9%
Rapid prototyping	10.2%	9.7%	n/a	n/a
Distance learning	5.6%	19.3%	n/a	n/a
ISO 14000 certification	1.1%	16.6%	n/a	n/a

Source: Youtie and Shapira, 1997.

TABLE 5-2 Use of Technologies and Techniques by Facility
Employment Size, 1996

Technologies and Techniques	All Respondents	Number of Employees		
		10 to 49	50 to 499	500+
PCs, non-manufacturing	96%	94%	98%	100%
JIT customers	58%	52%	64%	76%
MRP II	58%	51%	66%	85%
Preventive maintenance	53%	49%	57%	77%
LANs	51%	40%	63%	79%
Employee teams	49%	36%	63%	90%
PCs, shop floor	45%	32%	57%	88%
CAD, CAE	44%	35%	51%	85%
Internet	38%	38%	37%	55%
SPC, SQC	38%	22%	52%	87%
Electronic business transactions	37%	29%	46%	64%
NC, CNC	29%	25%	33%	34%
Data collection devices	27%	12%	42%	76%
CAD; CAM	23%	19%	25%	48%
Manufacturing cells	19%	13%	26%	34%
CIM	16%	11%	21%	40%
Automated material handling	15%	9%	21%	49%
ISO 9000/QS 9000 certification	14%	7%	20%	40%
Automated in-process inspection	11%	6%	14%	42%
Rapid prototyping	10%	9%	12%	13%
Distance learning	5%	2%	7%	26%
ISO 14000 certification	1%	0%	1%	8%

Source: Youtie and Shapira, 1997.

reported greater SPC, CAD, CAM, MRP, ISO, and HAZMAT capabilities.
Small SMEs considered improvements in payment terms and financing
somewhat more important than did large SMEs.

There were few differences between SMEs as a function of concentration of top customers, although a distinct difference was found in the
percentage of electronic transactions. SMEs with higher concentrations of
top customers had a higher percentage of electronic transactions (15 percent) than those with lower concentrations (7 percent).

GEORGIA TECH ECONOMIC
DEVELOPMENT INSTITUTE SURVEYS

Studies conducted in 1994 and 1996 by the Georgia Manufacturing Extension Alliance, through the Georgia Tech Economic Development Institute, confirmed many of the findings of the committee's survey. The Georgia Tech studies surveyed the manufacturing needs, practices, and performance of all manufacturing firms in Georgia with 10 or more employees. The approximately 1,000 responses are summarized in Tables 5-1 and 5-2.

Most of the technologies and techniques in the Georgia surveys are identified elsewhere in this report as becoming increasingly important for successful participation in integrated supply chains.

Some of the findings of all three surveys were confirmed by a survey reported in *The Wall Street Journal*, which showed that in 1999 only half of small businesses (defined as having 10 or fewer employees) had Internet access, and approximately 20 percent had their own Web sites. Another survey in 1999 showed that 54 percent of small businesses and 62 percent of medium-sized businesses (defined as having at least 100 employees) had some corporate Web presence (*Wall Street Journal*, August 17, 1999).

Finding. Survey data shows that the manufacturing world is changing rapidly and that small SMEs are significantly behind their larger counterparts in advanced technical capabilities.

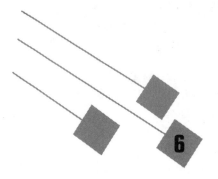

Quality, Cost, Service, and Delivery

Competitive quality, cost, service, and delivery have always been fundamental requirements of suppliers. They are still the cornerstones of integrated supply chain requirements for SME participation, although some aspects of these requirements are changing as integration levels increase.

QUALITY

Customer expectations are rising, and to remain competitive, OEMs are demanding higher quality from their suppliers. Automobiles last far longer than they did 20 years ago. Electronics, although they are orders of magnitude more complex than they were a decade ago (which should result in greater opportunities for failure), are far more reliable. The demand for six-sigma[1] and other quality initiatives is an emerging trend. The aerospace industry, among others, will almost certainly require improvements in supply chain quality as OEMs and prime contractors work towards the goal of producing defect-free work on the first try.

It is a fundamental premise of manufacturing that high-quality end products cannot be built cost effectively from low-quality components. Most suppliers operate in a tolerance range of two to three sigma. OEMs cannot achieve six-sigma quality with three-sigma suppliers. Suppliers

---

[1]Sigma is a statistical measure of the capability of a business or manufacturing process to perform defect-free work. The common measurement index is defects per unit. A unit can be virtually anything (e.g., a component, a line of code, or a customer invoice). At the six-sigma level, the incidence of defects is nearly zero (Velocci, 1998).

with quality deficiencies weaken the entire chain and, unless they improve, are vulnerable to being phased out.

Supply chain integration requires that quality be more than a set of abstract standards. Quality must be a systemic way of doing business that is instilled in all participants in the chain. Quality has become critical in supply chains using just-in-time manufacturing with low inventory levels because they have very few buffers to protect against quality failures.

SMEs should not consider quality only as a requirement for continued supply chain participation, but as a strategic capability. SMEs that adopt quality as a competitive strategy are finding that they are better able to weather cyclical swings in their businesses and that their product costs are lower. Thus, SMEs may reap benefits by exceeding the quality levels required by supply chains.

Most integrated supply chains require that participants have a carefully reasoned and executed quality plan that includes concerted efforts to provide levels of quality appropriate to the market being served. Proficient problem identification and problem solving capabilities are fundamental elements of the quality plan. Although six sigma and other quality programs may be of strategic benefit, they can be expensive to implement. Thus, SMEs must carefully target and prioritize improvements in terms of their effect on the company's operational and financial goals, as well as overall business objectives. Delivering a quality product requires, at a minimum, well established and well documented manufacturing processes and controls that meet impartial standards and customer requirements. Six-sigma is one such standard, but other, less exacting standards may be adequate. SMEs are increasingly being required to identify, capture, analyze, and act on process data in conformance with SPC. Many of these requirements are based on quality standards, such as ISO-9000, QS-9000, and ISO-14000.

SMEs should discuss with their supply chain partners how quality improvements can affect the overall performance of the supply chain. Together, the partners should identify and prioritize SME actions that will have the greatest impact on overall supply chain quality, cost, and cycle time and determine how these actions will translate into increased competitiveness and profitability for the SME.

Properly implemented quality procedures can reduce rework, scrap, testing, and inspection and improve on-time deliveries. The result can be substantial savings and fewer schedule variances. For example, in the development and pilot production phases of new electronic products, two new quality techniques, highly-accelerated life testing (HALT) and highly-accelerated stress screening (HASS) have yielded substantial benefits. Although they are somewhat expensive, these techniques have been shown to be effective in debugging new products and identifying weaknesses that could lead to operational failures. In many cases, these new techniques have been better able to identify problems in advance of full-scale production than previous methods, including MIL Standard tests. Used in conjunction with other quality techniques, such as SPC, HALT and HASS can provide substantial returns on investment.

SPC has advanced beyond its early role as an after-the-fact application of statistics to production and inspection data, when it served primarily as a means of creating a report verifying compliance with customer requirements. Today, SPC can provide opportunities for real-time assessment of manufacturing processes and can enable response to the causes of process variations as they happen. Thus, processes can be adjusted before more nonconforming products are produced. The savings are immediate and quantifiable, not just in direct costs, but also in more timely shipments, improved product quality, and increased customer satisfaction, all of which reflect favorably on SMEs seeking long-term supply chain partnerships.

Participants in a supply chain need a common language to facilitate accurate communication on issues of quality. Because such languages are not universally defined and can vary from chain to chain, quality standards, such as ISO, can be helpful. SMEs may wish to adopt such standards voluntarily.

SME participation in integrated supply chains can facilitate quality improvements through the exchange of "best practices" among partners, which can enhance understanding and provide examples of proven techniques. More advanced participants in the chain can assist those who are less advanced to adopt and use appropriate quality techniques.

Recommendation. In response to the requirements of integrated supply chains for improved quality, small and medium-sized manufacturing enterprises should adopt quality as a competitive strategy and consider implementing techniques, such as six sigma, ISO certification, and statistical process controls, to comply with customer demands, improve overall business performance, and provide a common language for communication on quality issues.

COST AND VALUE

Global bidding on the Internet has forced suppliers in many industries to slash prices dramatically. Costs have always been critical, and in the increasingly global economy it is not unusual for SMEs to find sudden gaps between their prices and the prices of competitors from low-cost areas. The convergence of (1) improvements in high-speed communications, (2) reduced transportation costs, (3) widespread adoption of

English as the language of business, and (4) universal access to technology and effective management practices has enabled companies in areas with low labor costs to become competitive regardless of location. Thus, many SMEs must substantially reduce costs to remain competitive, and they are finding that competing on the basis of cost alone is becoming a losing game.

Case Study: On-Line Commodity Buying

In some industries, geographic proximity is no longer an advantage. The Internet and modern transportation capabilities have combined to enable on-line businesses with low labor costs and appropriate capabilities to compete from anywhere in the world. These capabilities have eliminated two traditional advantages of local suppliers: their physical proximity and customer ignorance of comparison prices.

Large OEMs, including the Boeing Company and United Technologies Corporation (UTC), are taking advantage of these trends by turning to on-line bidding for the procurement of low-technology, pre-engineered items, such as nuts, bolts, and steel shafts, for which there are a large number of suppliers. Several companies have sprung up to conduct on-line auctions that pair worldwide buyers and sellers. FreeMarkets, for example, which conducts structured bidding events for industrial products, brokered more then $500 million worth of goods in 1998 and expects to handle three times as much in 1999. In a similar manner, MetalSite, conducts steel auctions; FastParts, Inc., auctions computer components; Inventory Locator Service LP auctions airplane parts; and Affiliated Networks, Inc., auctions marine supplies and boats. Although price is important, buyers may consider other factors in their final decision or may reject all bids. New suppliers are required to demonstrate appropriate capabilities prior to bidding. Unless the bids of new suppliers are substantially below those of incumbent suppliers, the jobs may go to incumbents because the buyer is more familiar with their capabilities or wants to retain a small base of the most competent suppliers.

In 1999, for example, FreeMarkets structured a daylong bidding event for UTC that included numerous lots of simple machined metal parts. Prior to the bidding, FreeMarkets analyzed a list of preapproved suppliers and selected ones acceptable to the buyer. The buyer was able to specify preferences, such as the inclusion of small and disadvantaged businesses. Each supplier was sent a package in advance detailing the parts being sought, pertinent quality requirements, and delivery dates. In this example, the bidding was conducted on a secure network. Suppliers were not informed of the names of their competitors or the prices paid for similar items in the past. They could, however, see rival bids in real time.

Bidding was described as "bare-knuckled," with low bids coming from qualified suppliers in India and elsewhere. UTC purchased $7 million worth of goods that day; savings totaled 25 percent. In one lot, an incumbent supplier was forced to reduce its price by more than 50 percent from its previous contract to retain the business.

These and other auction services are opening new worldwide markets to SMEs that have competitive costs and capabilities. Once they have become qualified suppliers, only a PC and a modem are required to participate. However, Internet bidding can dramatically reduce the profit margins of SMEs that have not properly positioned themselves in terms of cost, product differentiation, or added value. Because few SMEs have sufficient margins to withstand such competition, it is essential for their survival that they prepare in advance for such eventualities.

COST REDUCTION

Opportunities for internal cost reductions include direct labor, materials, scrap, and rework. Other opportunities can be found through creative reductions in overhead. The Boeing Company, for example, reports saving hundreds of millions of dollars by moving its system of spare parts sales and aircraft maintenance manuals to the Internet. Techniques, such as just-in-time manufacturing, activity-based costing (ABC), vendor-managed inventory, and lean manufacturing, can be used effectively to reduce non-value-added costs.

SMEs may find ABC especially helpful. Traditional accounting systems accumulate costs, such as engineering and material handling, into overhead accounts, which are then allocated to products based on the amount of direct labor each product requires. This approach is useful when there are long manufacturing runs and direct labor is a large part of total costs. However, traditional accounting systems are less useful for firms with substantial investments in information, product design, and agile manufacturing technologies. ABC assigns job costs based on the actual use of resources, enabling firms to price their products appropriately, determine in which markets they can compete effectively, make better capital allocation decisions, and calculate the incremental costs associated with potential courses of action.

The implementation of ABC requires the following steps:

- creation of a conceptual outline of the firm's cost-flow patterns
- development of a day-to-day activity-based cost accounting system that collects the costs associated with each product or process
- construction of a computer-based cost accumulation model that simulates the activity-based cost flows

- analysis of the differences between the new and old systems to validate the usefulness of the new system for gathering accurate job costs

Just as OEMs use outsourcing, SMEs must consider creating and integrating their own supply chains to optimize their own cost structures. The benefits available to OEMs from integrating supply chains are also available (on a smaller scale) to SMEs. This potential for cost improvement has yet to be exploited by most SMEs. Participants may, for instance, reallocate work among themselves to improve the overall efficiency of the supply chain. Several personal computer companies have implemented "channel assembly," delegating responsibility for final assembly to distributors with specific customer knowledge and lower labor rates. This practice can be used to reduce inventory levels and the probability of obsolete inventories. Other supply chains have succeeded in redeploying, consolidating, or sharing warehouse space and inventories among participants to reduce overall costs to the chain. Participants in integrated supply chains may also be able to share (or preferably eliminate) some administrative procedures.

Although cost reductions are critical, additional efforts will be required to maintain competitive advantage and meet the challenge of increasingly competent global competition.

ADDED VALUE

Many SMEs will have to do more than provide low-cost parts if they want to become partners with demanding customers. They may have to simultaneously maximize the value and minimize the total cost of the goods and services they provide. Competitive advantage can be achieved through value-added services, including low-cost storage, rapid response in dealing with warranty issues, ready access to spare parts, and improved logistics. Investment in enhanced product design capabilities can create other opportunities for adding value, linking an SME more closely to the OEM. SMEs should also investigate other innovative opportunities, such as building subassemblies instead of just parts.

Although integrated supply chains are increasingly recognizing the benefits of added value, some customers may be unwilling to pay for it. Thus, SMEs may have to reposition themselves in new industries and find new customers that are willing to pay for value-added products and services. SMEs must carefully identify customer preferences, buying habits, and unfulfilled needs and determine whether efforts to meet them would be appropriately rewarded.

Recommendation. Small and medium-sized manufacturing enterprises should rigorously reduce costs internally and throughout their supply chains. They should also seek ways to increase the value added to their products and services and find customers willing to reward such value.

DELIVERY

As levels of supply chain integration have increased and inventory levels have been reduced, reliable, on-time deliveries have become increasingly critical for success. Large inventories and production capacities were traditionally required to ensure on-time delivery. However, with advanced information systems, deregulation, agile manufacturing organizations with flexible equipment and tooling, and sophisticated logistics systems, integrated supply chains no longer need large, costly inventory buffers to respond to unexpected events and variations in demand.

Manufacturing supply chains can benefit from integration with logistics providers/carriers, many of whom offer services that can reduce supply chain costs and increase overall performance. Reduced transportation costs, shorter in-transit times, and value-added services can be major factors in improving the competitive position of the supply chain.

For example, by outsourcing logistics to the FDX Corporation, Vishay Intertechnology has reduced cycle times from factory to customer from as long as 12 days to 2 days (Tanzer, 1999). At a FedEx warehouse in Subic Bay, the Philippines, $1-per-hour labor picks and packs components manufactured in Taiwan, China, and the Philippines for same-day shipment to customers around the world. The international air express industry is a key enabler of many trends in today's global manufacturing, including shorter product cycles, e-commerce, just-in-time manufacturing, mass customization, reduced inventories (especially of high-technology products that can rapidly become obsolete), and global sourcing and selling. Air freight is increasingly being used, not just to deliver electronics, but also to deliver pharmaceuticals, medical equipment, automotive parts, designer clothing, perishable agricultural products, and replacement parts of all kinds.

Both FedEx and DHL International have made extensive investments, not only in logistics and information systems, but also in parts centers and assembly operations. For instance, in Brussels, DHL performs upgrading, repair, and final configuration of computers for Fujitsu and Stratus, projectors for InFocus, and medical equipment for Johnson & Johnson. DHL also stores and supplies parts for EMC and Hewlett-Packard and replaces mobile phones for Philips and Nokia.

Many OEMs have reduced their own inventories by foisting

inventory ownership on suppliers, leaving them with the physical and finåncial responsibility for maintaining inventories. Unless suppliers can perform these services at lower costs and are appropriately compensated, this practice only moves costs around and potentially weakens suppliers. Improved communications and coordinated materials and capacity planning are required for the cost-effective reduction of inventories, the elimination of bottlenecks and idle capacities, and matching of capacity to demand. To participate at this level of planning, an SME must be able to articulate capacity plans and strategies and understand their implications up and down the supply chain.

Worldwide transportation costs can be low compared to costs associated with inventories and labor. For instance, suppliers in regions with low labor costs routinely ship heavy rolls of steel all over the world. Although transportation can provide added value and competitive advantage, the costs and benefits of using advanced transportation methods must be monitored carefully. Reliance on overnight deliveries can be an expensive alternative to on-time production. Some companies are finding that the transportation costs required to enable low inventories and just-in-time manufacturing exceed the savings. Thus, SMEs should attempt to implement the "right" logistics strategies to minimize overall costs.

Recommendation. In response to increasing demands for rapid delivery and customized products, small and medium-sized manufacturing enterprises should consider using advanced supply chain communication systems, flexible manufacturing techniques, and modern transportation capabilities as alternatives to investing in large inventories and production capacities.

SERVICE

Customer expectations for timely service before, during, and after a sale continue to increase and, aided by the Internet and modern transportation methods, suppliers are responding to these demands. Web sites are being used to post all manner of nonproprietary information, including maintenance manuals, service bulletins, and responses to frequently asked questions. e-Commerce enables customers to place orders around the clock from anywhere in the world without incurring the cost of long-distance calls. Replacement parts can be delivered overnight in the United States and within a few days almost anywhere else.

In February 1999, Boeing executives warned Graybar Electric, Inc., that it would soon eliminate suppliers that did not have a Web presence. In March, Motorola warned Graybar that suppliers that did not implement Web-based commerce within the next year would probably be locked out of their business for good (Royal, 1999). Graybar introduced its extranet just in time to meet customer demands. For some businesses, however, going on line may not provide a competitive advantage. Phoning a local restaurant for carryout is still quicker and simpler than ordering carryout on the Internet.

Recommendation. In response to increasing customer expectations for service and support, small and medium-sized manufacturing enterprises (SMEs) should reassess their service and support capabilities and revise them, as needed, to remain competitive and to seize new market opportunities. SMEs should develop an understanding of the opportunities provided by various Web technologies and, if appropriate, create a Web presence.

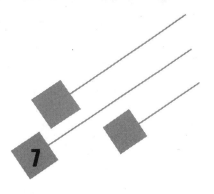

Building Partnerships

Companies worldwide are creating alliances that are reshaping the basis of competition. Remaining competitive in these changing business environments requires the ability to create appropriate types of partnerships as needed. *Partners* may be defined as companies that agree to work together, often for a specific period of time or to achieve specific objectives, and share the risks and rewards of their relationship. In practice, partnerships are defined by both a relationship agreement between the partners and by their day-to-day interactions.

There are many types of partnerships including, for instance, joint ventures, licenses, and cooperative marketing and research agreements. Each party must carefully analyze its strategic needs and the risks and benefits of various types of partnerships before entering into an agreement. To reduce misunderstandings, participants should collectively agree on objectives, definitions of success, methods for sharing risks and rewards, and metrics for measuring performance. To succeed, partnerships must be based on trust and respect or, in rare cases, on an overriding need to achieve shared objectives.

Integrated supply chain partnerships involve more than legal agreements to work together. They involve complex interrelationships of shared activities, processes, interests, objectives, and competitive information. Information sharing and extensive cooperation are necessary for partners to react effectively to changes in the marketplace.

The participants themselves, who often have different views depending on their corporate objectives and roles in the supply chain, must determine the specific configurations of the partnerships. The creation of successful partnerships requires visionary leadership, understanding of supply chain roles, and interpersonal and legal skills. Partners should consult with an attorney to avoid violating laws against restricting competition. Creativity and sensitivity to relationship issues are critical skills for overcoming problems that arise in intimately coupled, highly integrated supply chains. Only a constructive attitude, visionary leadership, and a strong commitment to mutual success can enable the highest levels of supply chain integration.

CONTRACTS

The underlying mechanism of supply chain integration is the development of strong customer and supplier relationships based on mutually agreed-upon performance standards. These relationships are usually defined by means of a contract, the proper structure of which is important for successful integration. Integrated supply chain relationships are often defined by additional agreements regarding product and financial flows, channel (product distribution) policies, price protection, contingencies, and capacity reservations (i.e., the guaranteed availability of specified levels of production capacity). Unlike purchase orders, which tend to be one-sided and generally ignore the issues of risk sharing and common goals, contracts are structured to meet the needs of the partners. Before an appropriate contract can be crafted, the partners should agree on a shared vision, objectives, and process framework. (It should also be noted that although contracts are recommended, they are not required and it is possible to achieve effective supply chain integration without them.)

Several aspects of contract development should be considered: (1) the difference between a myopic and a farsighted view of contracting; (2) the distinction between a contract structured as a set of legal rules and a contract that serves as a framework for the relationship among partners; (3) conflict avoidance and resolution, mutuality, creation and maintenance of order, and alignment of risks and rewards; and (4) different ways of structuring relationships among supply chain participants.

Farsighted Contracting

Some contracts are simple, such as those for the procurement of standard items, like office supplies, which are usually kept in stock and for which there are many suppliers. Unless large orders must be filled, these procurements do not require advanced planning or the creation of custom contracts. If specialty items, nonstandard tolerances, or special deliveries are involved, the nature of the contract and its capacity to serve the needs of the parties becomes more important. Each party will want to look ahead, anticipate potential problems or hazards, and find ways to relieve these conditions by providing for them in the contract (or, more generally, in the overall supply chain relationship).

High levels of supply chain integration are more likely to be achieved with "win/win" contract approaches than with win-lose approaches. A long-term view is important for integrated supply chains in which the parties have a substantial bilateral stake in the relationship and contractual breakdowns can be costly. The typical intent of a partnership is to establish a relationship with increasing interdependency that will outlive the term of the contract. A farsighted view, although it may be difficult to implement, can increase benefits for all parties.

Contract as a Framework

Contracts consisting of rules and contracts that serve as an operating framework for a relationship are very different. Lawyers and accountants often relate to the former, whereas business is more often conducted in the spirit of the latter. Contracts that serve as frameworks for relationships are based on four key ideas: (1) the objective of the contract is to serve the goals and desires of the parties, rather than those of the lawyers; (2) contracts that attempt to define complex relationships are unavoidably incomplete; (3) give and take by all parties is necessary to work through gaps, errors, and unanticipated situations; and (4) informal and formal features of the contract and of the organization are all part of the exercise. This is not to say that the letter of the contract is unimportant. If, despite their best efforts, the parties are unable to work through a contractual impasse, the contract can be useful for purposes of ultimate appeal.

There is every reason, therefore, for contracts to be written and negotiated carefully, although this does not mean being legalistic. Because providing for every possible contingency is impossible, information disclosure and adaptive mechanisms must be included to assist the parties when difficulties arise and the formal terms of the contract take on added importance. Contracts are legal documents, after all, and relationships should be defined precisely in case the contract becomes the ultimate reference for resolving disputes. Mainly, however, contracts should be thought of as frameworks to aid the parties in realizing their collective purposes and mutual gains.

Neither party, especially not an SME, can unilaterally decide that a framework is the correct approach to define a supply chain relationship. Both parties must subscribe to the concept for it to be successful. If one party adopts a myopic, legalistic view and the other views the contract as a framework, both parties will be frustrated. It is important, therefore, that they agree on the nature of the relationship and recognize that each party may have a different bottom line. Incomplete contracts can be fraught with hazards because the best short-term interests of one party rarely coincide with the short-term interests of the other. Therefore, a long-term view of the relationship is important, and a contract constructed as a framework helps to accomplish this.

An ideal supply chain contract is based on a history of mutual trust and provides a "governance" structure that allows both parties to operate in a flexible, yet disciplined way, mitigates conflict, and aligns and shares both risks and rewards. Developing mutual experience and trust takes time, and a contract that outlines only the framework of a relationship may be impossible without it. Therefore, contracting should be considered an evolving process that proceeds in a slow and exploratory way, building on successes. Learning how to deal, adapt, and relate to each other is a normal part of a productive contractual exercise.

Appropriate Contracts

Supply chain relationships can be organized in many ways, and contracts must be customized to address the diversity of transactions and industries. Transactions, for instance, can be generic or specialized. The former are simple; the latter are often more complicated. Contracts typically address the size of the job in relation to the capabilities of the supplier, the length of the contract, the amount and kind of specialized investment required to support the transaction, the extent of supplier involvement in the design prosess, and the degree to which implementation entails real-time responsiveness. Contracts often contain security clauses addressing specialized investments and proprietary intellectual property. Specialized equipment that cannot be redeployed without the loss of productive value poses a greater risk than generic investments. Hence, added security features may be required in the contract to cover the redeployment of specialized equipment. In general, contract complexity increases with the magnitude of the job, the length of the contract, the amount of specialized investment, the degree of design investment, and the need for real-time responsiveness.

A basic procurement contract specifies a price, addresses contractual hazards, and provides for security features. If the hazards are great and the buyer refuses to safeguard the supplier against risk, the supplier should reflect this by charging a higher price (risk premium). However, considerations of mutuality suggest cost-effective security features be crafted so that the supplier is partially relieved of risk and the buyer pays a lower price.

Contracts regularly allocate risks and rewards. If the contract covers a single transaction of short duration, the parties typically negotiate hard to secure all of the benefits and to transfer all of the risk to the other party. However, if the agreement is inequitable, the long-term relationship, which is essential for supply chain integration, is bound to deteriorate. Incentive systems embodied in the contract should ideally align the parties so that each has an appropriate stake in the success of the joint undertaking. Generally, the party with the greatest influence on the success or failure of the venture should be allocated greater portions of the risks and rewards.

A legalistic way to safeguard transactions is to include financial penalties for a premature breach of the terms of the contract. For example, both parties might make specialized investments for which redeployability outside of the contract will be difficult. Thus, their exposure is symmetrical. Alternatively, the parties may provide for the reciprocal disclosure of information and auditing contingencies. The parties could also agree in advance to specialized mechanisms for settling disputes, such as arbitration. Contractual risk can also be reduced by ensuring that the staff members responsible for implementing the contract have backgrounds in and an appreciation of the continuity value of the relationship.

Finally, differences among industries with respect to maturity, technology, size distribution of firms, and product differentiation can pose different contracting and organizational challenges. In general, a mature industry with a dominant design and multiple firms producing homogeneous products poses fewer problems for contracting. For newer industries with rapidly changing technologies, in which a few firms are competing for the dominant design and in which real-time responsiveness and innovation are critical, contracting is more complicated.

Recent Changes

As supply chain participants increasingly form alliances and partnerships, critics are starting to worry about the impact on competition. Regulators discourage schemes that reduce consumer options or price competition. In October 1999, the Federal Trade Commission, in consultation with the U.S. Department of Justice, issued antitrust guidelines for collaborations among competitors. The committee suggests that SMEs obtain competent legal advice to avoid legal problems associated with these complex issues.

Recommendation. Small and medium-sized manufacturing enterprises should develop a basic understanding of partnership agreements and, with legal assistance, use partnerships as a means of improving their responses to changing business conditions.

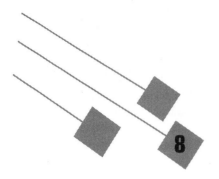

Management Skills and Human Factors

Managing an SME in an integrated supply chain is a complex task, and participation in multiple chains adds to the complexity. Rapid changes in the business environment, shorter product life cycles, and increasing customer demands require a robust management team that is willing and able to respond to changes. Not only have the demands for quality, cost, service, and delivery increased, but the range of required capabilities, techniques, and skills has grown more complex. Simultaneous with continuing day-to-day operations, managers of SMEs may be under pressure to remake their businesses to meet anticipated demands.

LEADERSHIP, VISION, AND STRATEGIC DIRECTION

Corporate management has traditionally focused primarily on the internal workings of the firm. However, supply chain integration and the changing global manufacturing environment now require that management devote more attention outside of the firm. More frequent interactions with customers and suppliers are important components in a strategy responsive to evolving business opportunities. Indeed, in a world of evolving opportunities and advantages, managing change and innovation is becoming a required core competency. Managers must be able to assess the changing business landscape, lead the enterprise in a visionary manner, and create and operate an evolving, integrated chain of appropriate capabilities both within and outside of the corporation. As the pace of business change accelerates, management must have the drive and determination to reinvent the company as necessary to meet changing needs.

For instance, a company may have to adapt its operations to achieve preferred supplier status when a customer reduces the number of supply chain participants. Then, as a surviving preferred supplier, the SME may have to assume increased strategic responsibility, accepting larger roles in product design, service, and coordination of the output from its own suppliers to provide seamless, total solutions to the OEM.

As the business world evolves at an increasing rate, no lead in capabilities lasts forever. Companies with extensive investments in today's technologies may be reluctant to replace them with new technologies. This hesitancy can provide a window of opportunity for others. Thus, incumbents must avoid becoming complacent, and outsiders must be alert for opportunities. At the same time, the tendency toward long-term relationships and joint investments makes it increasingly difficult for newcomers to unseat incumbents in an integrated chain. Some incumbent suppliers are assured of the business as long as they remain competitive. Because of the rapid rate of change in many industries, opportunities arise and disappear frequently, and taking advantage of these opportunities can be expensive. Therefore, opportunities must be carefully selected so as not to exhaust an SME's limited resources. Mapping of requirements and capabilities (see Chapter 12) can be useful for assessing and prioritizing opportunities.

Many SMEs do little formal business planning. A 1999 survey of 500 businesses with fewer than 500 employees revealed that only 13 percent have an annual budget in writing, 14 percent have an annual business plan in writing, and 12 percent have a long-range plan in writing. Many respondents indicated that in the "real world" life is simple: their annual plan is in their head and their annual budget consists essentially of "sell as much as possible, make payroll, and keep the lights on" (*Wall Street Journal*, September 7, 1999).

Developing a strategy or business plan need not be an elaborate process, but it should involve more than just an annual budget. At a minimum, the plan should define a vision for the company and identify specific steps and timetables for achieving the vision. It should also address each of the key topics identified in this report and present financial projections for the next several years. It should zero in on core competencies and outline plans for additional value-added services.

For small enterprises, the plan can focus on a small number of activities. An SME may be able to make a substantial profit by focusing on one critical product or process or by providing convenient, low-cost production capacity to meet surges in OEM demand. Therefore, despite the need to be aware of changes in technology and customer requirements, the optimum strategy for a small SME may be to excel in one carefully defined capability and retain merely adequate capabilities in other requirements of an integrated chain. The committee believes strongly in the value of business planning and suggests that SMEs that are unwilling or incapable of preparing business plans are unlikely to be successful in integrated supply chains.

Recommendation. Although extensive formal planning may not be justified, it is becoming imperative that small and medium-sized manufacturing enterprises periodically pause from the rush of daily business to survey their business environment of rapidly changing technologies and customer requirements and develop a brief, formal business plan.

SUPPLY CHAIN INTEGRATION

Few SMEs have availed themselves of the benefits of integrating their own supply chains, a core competency that can have a huge effect on their success or failure. Management expertise in forecasting markets and technological directions, selection of the right portfolio of capabilities, and correct decisions regarding internal investments and outsourcing can largely determine a company's competitive advantage. Despite the natural desire to retain capabilities in house, an SME must be brutally analytical in deciding (1) which capabilities are essential for its competitive edge or can be done at lower cost in house, and (2) which are noncore competencies that can be outsourced at lower cost as part of a well integrated supply chain. Retaining knowledge, for example, is important. But outsourcing surge production capacity and noncore capabilities, such as information technologies to lower cost providers can often reduce operating costs and allow an SME to concentrate on the remaining in-house operations. Internal development, acquisitions, or expansion of the capabilities of external segments of the supply chain can fill gaps in the corporate capability chain. The right approach depends on many factors, including the availability of capital and the need to retain control of capabilities that provide competitive advantage.

Barriers to Integration

As SMEs strive to integrate their own supply chains, they must be aware of several barriers that can impede the integration process. First, and foremost, they must overcome the tendency of participants to work for their own advantage with little regard for the effects on other participants or the supply chain as a whole. Optimization of one part of the supply chain frequently results in suboptimization of the entire system. Greater benefits can usually be attained for all participants by optimizing the system as a whole. Other integration challenges can include:

- lack of clearly defined, mutually acceptable goals
- cultural incompatibilities
- poor communications
- lack of clearly identified mutual benefits
- lack of in-depth commitment to and nurturing of relationships
- lack of metrics or inappropriate metrics for evaluating performance
- incompatibility of capabilities in communications and/or electronic design technologies
- lack of, or breakdown in, mutual trust and respect
- clashing corporate cultures
- legal barriers, including intellectual property, liability, and anti-trust issues
- government procurement policies and regulations

Forces beyond the control of supply chain partners can sometimes undermine the best efforts at integration. The ability to foresee these challenges, understand the capabilities and limitations of supply chain participants, and formulate effective responses are critical skills for supply chain managers. With these skills and an understanding of the dynamics and interrelationships of the supply chain, timely, effective responses to disruptions and imbalances can be implemented.

Supply chain effectiveness can always be improved. The challenge is to prioritize opportunities correctly, allocate scarce resources, and devote the time and expertise necessary to harvest these opportunities. SMEs that understand supply chain integration and carefully pursue the right opportunities can achieve a substantial competitive advantage.

MANAGEMENT SKILLS

Participation in integrated supply chains requires management skills that may be different from the command-and-control style of management found in many traditional SMEs. A change from the command-and-control style to a leadership-and-facilitation style can be difficult, especially in a privately owned SME. According to Bellman in *Getting Things Done when You Are NOT in Charge,* participants must learn to work in support of the goals of the integrated chain while simultaneously striving to realize the goals of the individual enterprise. At times, an SME may have to compromise aspects of its independence in the interest of "team play" and consensus. Moving from a command-and-control model to a model characterized by shared interests, shared resources, and shared risk may be uncomfortable for traditional managers and can require considerable time and effort (Bellman, 1993).

Managing in the "information age" also requires skills in guiding the informal human networks that develop in corporations and integrated supply chains. Much of the essential knowledge in an organization comes from hands-on experience and informal human networks that are embedded in the corporate hierarchy. Those who guide change must learn to harness the power of these networks, which are often in tension with the management hierarchy. According to Karen Stephenson of UCLA's Anderson School of Business, "hierarchies are about authority and are rigid, but networks are about how things actually get done and are thus adaptable and constantly shifting. If this mismatch becomes pronounced enough, the network can destroy the hierarchy" (*Washington Post*, March 6, 1999).

Decision-making capabilities are, in some ways, more critical in SMEs than in large companies. Large companies have greater financial depth and a wider range of business opportunities; a wrong decision (e.g., customer selection, equipment investment, or technology acquisition) will not cripple the enterprise. SMEs have fewer resources and can afford fewer mistakes. Committing a significant portion of an SME's resources (i.e., people, equipment, and financial resources) to one customer or project can preclude participation in others.

RISK AND INNOVATION

To succeed in the changing supply chain environment, SME management must be willing to take carefully calculated risks to promote innovation and maintain their company's competitive advantage. Paradoxically, trying to maintain the status quo and avoiding risk is, in itself, a risky strategy. Pursuing a diverse customer base can help an SME mitigate some business risk, although the added customers may create additional and conflicting demands. Thus, an SME must aggressively eliminate bad customers, pursue good ones, and improve relationships by building and nurturing trust. For instance, it may be advantageous for an SME to be tightly linked to one key customer or supply chain for a while, even at the risk of becoming essentially a captive. This approach, although risky and seldom recommended, may actually constitute a lower risk than trying to invest to meet the conflicting demands of multiple customers.

SMEs are often assumed to have a significant advantage in their ability to tolerate "radical" new ideas and to nurture innovators rather than smother them with bureaucracy, tradition, and the status quo. Peter Drucker, writing in *Forbes*, supports the importance of an innovative environment, emphasizing that management's job includes creating and managing an innovative environment. "A management that does not learn to innovate will not last long" (Drucker, 1998).

HUMAN FACTORS AND SKILLS

Barriers to effective participation in supply chains often have more to do with cultural and interpersonal issues than technical issues. Successful participation in integrated supply chains requires a high level of interpersonal skills. As is well known in the business community, long-term personal relationships between customers and suppliers are important in building a stable and growing business. Person-to-person relationships help to form bonds of trust that enhance credibility on both sides. Among the most important relationships are those between day-to-day decision-making peers at all levels of the customer and supplier organizations. Design engineers, sales executives, and logistics experts, to name a few, from each of the key participants should be working together to increase the extent of integration and mutually increase the efficiency of their processes and functions. Successful integration requires tight, cohesive relationships based on common goals between individuals at many levels and in many functions throughout the supply chain. For example, supplier manufacturing executives should develop strong relationships with the customer's procurement executives to ensure that they thoroughly understand each other's needs and capabilities. This understanding is crucial for effective decision making, meeting customer needs, and balancing available resources.

In another aspect of supply chain integration, it is becoming increasingly common for OEMs to give their assembly-line operators the authority to release shipments from suppliers, allowing the operators to interact directly with their counterparts in the supplier company instead of through the formal organizations. To ensure quality, the people responsible for making parts are sometimes sent to customer facilities to gain an understanding of the customer's problems and needs. All of these interactions require trained employees who can carry out these responsibilities efficiently and constructively.

Supply chain integration and the implementation of new technologies must be planned and handled with care because all of the participants in a supply chain are independent entities. Participants tend to resist changing their work practices, especially if the benefits are perceived to be greater for other participants. These difficulties can be compounded by the human tendency to resist technologies that cannot be easily implemented. Societal beliefs in individualism and autonomy generally run contrary to the ideas of information sharing, openness, win-win negotiations, and sharing of risks and rewards, all of which are important aspects of successful integration (Baba et al., 1996). Thus, considerable leadership and people skills are necessary to achieve successful supply chain integration.

Companies that own intellectual property will naturally resist sharing their hard-earned knowledge and skills unless they are reasonably confident that (1) they can trust the other party not to steal their secrets, and (2) they have enough clout to make it highly unlikely that a theft will occur. Yet the drive toward increased supply chain integration is heavily dependent on openness. Thus, participants must find ways to craft relationships and agreements that effectively guard against the misappropriation of intellectual property.

Similar issues arise when OEMs demand detailed cost structure information from privately held suppliers. Although some of this proprietary information can be valuable for collective efforts to increase supply chain efficiency and reduce costs, privately held SMEs are justified in releasing only selected information pertinent to specific supply chain initiatives.

Successful partnering requires a shift from the traditional attitude of "what's in it for me" to a new attitude of "how can we maximize our common good and what can I do to help us achieve our mutual goals." This change in philosophy can be extremely difficult and cannot be created merely by the negotiation of a legal contract. It requires trust and a spirit of giving, both of which are difficult to create without extensive interactions between the partners over time.

Although supply chain integration often focuses on the implementation of compatible technological systems, in some cases technology can aggravate the development of trusting, cooperative relationships. At times, human issues far outweigh technological issues. Thus, progress in supply chain integration requires that improvements in human interrelationships keep pace with attempts to achieve technological integration.

SUPPORT

Many OEMs demand that participants in their integrated supply chains upgrade their support functions. Each supply chain and each industry require somewhat different support capabilities. Defense industry suppliers, for example, must have capabilities in contracting and dealing with FAR. Demands for improvements in safety and environmental capabilities are being driven partly by government regulations and partly by good business sense. Integrated supply chains practicing just-in-time manufacturing have little slack to accommodate safety or environmental incidents.

LEARNING AND REDIRECTION

The competitiveness of SMEs in supply chains is becoming increasingly dependent on workers who can assimilate large amounts of information and use it to make effective decisions. "Virtual manufacturing facilities" will require that operators be capable of learning new skills in a changing environment. Supervisors and plant managers will need extensive systems knowledge in addition to a thorough understanding of processes and customers. Executives, to be effective in leading and redirecting the organization, must understand changing customer requirements as well as new technologies.

The wealth of information on the Worldwide Web is virtually useless without reading and learning skills. Turning raw data and information into actionable intelligence and knowledge is a critical skill for business leaders. SMEs that are limited in their ability to learn and reshape their businesses in the face of changing environments will not be able to sustain their competitive advantage. Indeed, learning should be treated as a core competency that can be used for competitive advantage. Thus, SMEs must develop organizations that are willing and able to learn and adapt.

Recommendation. Small and medium-sized manufacturing enterprises should (1) assess and strengthen their management capabilities; (2) create a corporate environment conducive to the flexibility, change, evolving skills, and learning required by integrated supply chains; (3) integrate their own supply chains; (4) learn to deal effectively with risk; (5) develop the people skills required to integrate effectively with customer supply chains; and (6) engender a shift in corporate attitudes about supply chains from "what's in it for me" to "how can we maximize the common good." Because all of the requisite skills are rarely resident in a single entrepreneur, whenever possible, SMEs should increase the breadth and depth of their management teams.

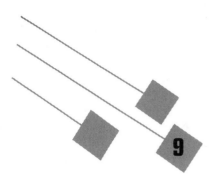

Technology

Technology is playing an increasingly critical role in the success or failure of SMEs. Computerized machines are replacing manual machine tools, CAD is replacing manual drafting, and computers are being used to track inventories, even in small shops. Although up-to-date manufacturing and process technologies are critical, they are no longer the only required technologies. Information technology has become one of the keys to operating success. Internet technologies alone are changing the mechanisms of communication, marketing, selling, buying, and generating revenue.

Suppliers are finding that one of the few escapes from the relentless pressure to reduce prices lies in change and innovation. The addition of value through innovative product and process design can sometimes differentiate the output of an SME from its competitors enough to enable profitable operation even in areas with high labor costs, such as the United States.

Successful companies distinguish themselves from their competitors by anticipating opportunities, selecting appropriate technologies, and using them for competitive advantage. Many SMEs, however, find gaps between customer supply chain requirements and their technological capabilities. The following sections offer suggestions for addressing some of these gaps.

ELECTRONIC COMMUNICATIONS, INFORMATION TECHNOLOGY, AND e-BUSINESS

The advent of the Internet and e-business is increasing competition by facilitating comparison shopping, raising customer expectations, providing ready availability of customized products with expanded features, and reducing costs. The Internet is a powerful and effective integration tool that can enable substantial improvements across the supply chain. It offers SMEs, in their roles as suppliers, manufacturers, and customers, a huge potential for exchanging information easily and securely with other supply chain participants. According to J. Cohen, the Internet can save supply chain participants from 18 percent to 45 percent in logistics costs through quicker order placement, faster delivery of goods, fewer transaction errors, and more accurate pricing (Cohen, 1999). The Internet can also be used to help manage inventory by providing a framework for just-in-time manufacturing. Inventory reductions during the past 10 years were enabled primarily by expensive electronic data interchange systems that linked inventory databases. Only wealthy companies could afford these systems. Now, versatile applications of the Web are providing similar opportunities for SMEs at substantially lower costs. The Internet can also be useful for removing traditional geographic barriers to collaboration and integration within the supply chain.

By using the Internet as an integration tool, SMEs can

- improve reaction to changing demands and markets
- optimize resources throughout the supply chain
- more efficiently source lower cost materials
- achieve shorter lead times and better due-date performance

Communications

Although face-to-face contacts and the use of telephone, fax, and surface mail are still essential, they are no longer sufficient to ensure competitiveness. Effective communication requires additional capabilities, such as e-mail, electronic data transfer, and more. Even small suppliers should consider creating a supporting information infrastructure of appropriate size and complexity to provide information in useful form to the right people at the right time. New communications technologies can provide substantial benefits to members of the supply chain. For example, traditional methods of processing purchase orders are slow and costly. The OEM receives a customer order, enters it into the MRP (materials requirements planning) system, writes its own purchase orders in response to component forecasts from the MRP system, and sends them to suppliers. Upon receiving the orders, suppliers execute the same procedures in their companies, and so on throughout the supply chain. With current electronic communications technology, purchase orders for lower tier parts and services can flow almost instantaneously and at virtually no cost throughout the supply chain. Although up-front investments for these electronic systems can be significant, they reduce the administrative costs of placing orders and can dramatically improve lead times and responsiveness.

There are many levels of data and communications integration. An SME, commensurate with its resources, should determine the appropriate level. At a minimum, the basic capabilities should include electronic networks based on generally accepted data transmission protocols, such as e-mail and file data transfer on the Internet, and private couriers, such as FedEx. No SME can expect to remain competitive without all of these. The following statistics provide examples of the ubiquity of these technologies (Ferguson, 1999):

- A billion e-mails were sent in 1998.
- Worldwide fax transmission minutes increased from 255 billion in 1995 to 395 billion in 1998 and will continue to grow rapidly for the next several years.
- Faxes account for one-third of phone bills at large corporations.
- With lower phone rates and better equipment, the average cost of sending a three-page long-distance fax dropped from $1.89 in 1990 to $0.92 in 1999 via stand-alone machine and to $0.30 via PC.

All levels of integration require the use of a common syntax and semantics so that cross-organizational data can be interpreted identically by all participants in the supply chain. Integration at the highest levels may involve object-oriented data modeling, data warehousing, high-level data protocols, and knowledge-based systems to provide almost instantaneous sharing of knowledge. Implementation of these systems requires extensive time and capital, as well as cultural changes associated with the transition from independent tools used by individuals to dependent tools that link people and organizations throughout the chain.

Recommendation. Although the highest levels of communication capabilities can provide incredible competitive power, they are too complex and costly for most small and medium-sized manufacturing enterprises (SMEs). These technologies should be monitored closely, however, because their costs and ease of implementation are improving dramatically. Internet technologies can provide many of these capabilities today at far lower cost, and SMEs should take advantage of these easy-to-use technologies.

Data Collection and Information Management

A research project by the Cambridge Information Network involving more than 270 corporate chief information officers found that 56 percent plan to install an Internet-enabled real-time link from their enterprise resource planning systems to their suppliers. The market for supply chain management hardware, software, and services is expected to increase by 50 percent in 1999 to $3.9 billion (Wreden, 1999). However, with all of the publicity about interconnectivity, it is easy to lose sight of the essentials of data collection and information management. First and foremost, SMEs must decide which data will be collected and how they will be used. The data management system should then be selected using technologies compatible with other supply chain participants. Competitive advantage is not gained simply through faster communication of data but from the skilled use of the knowledge gained from useful and timely data.

Decision making can be improved through real-time knowledge of sales rates, inventory levels, and production rates throughout the supply chain. These data can be used to reduce oscillations in the system (more closely matching production and inventories to demand) and reduce supply chain misalignment (Figure 9-1). Misalignment occurs when product shipments plus channel fill (the amount of product in the distribution system) are not equal to customer demand. Integration of demand planning, order fulfillment, and capacity planning can enable reduced supply chain inventory levels, improve on-time delivery, and enable more rapid resource deployment to problem areas. By minimizing multiple tiers of report writing and other non-value-added activities, overhead costs and delay times can be greatly reduced.

Collaborative demand planning based on shared customer (order and point of sale) and operational (inventory and availability) data can provide real-time forecasts of demand for all of the partners. If everyone in the supply chain has an immediate and accurate view of demand, the response of the chain, as a whole, can be rapid and synchronized. Distorted or delayed demand forecasts can cause inventory imbalances, delays, and higher costs. Accurate predictions require that SMEs have accurate and systematic methods of generating and sharing demand and supply information.

Several problems with information technology can hinder integration and planning. Information from various functions may be handled by multiple (and sometimes incompatible) electronic systems. Most organizations continually upgrade and migrate their software making it difficult to unite applications in the integration process. Information and

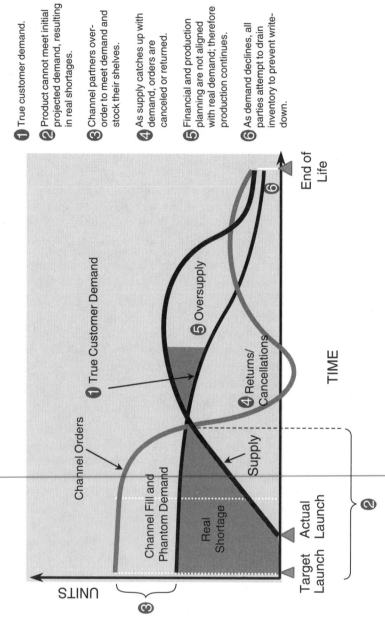

FIGURE 9-1 Effects of misalignment in the supply chain. Source: Murphy-Hoye, 1999.

functions are often scattered across the enterprise. Customer orders may be received through a sales force, while finished goods, inventory levels, purchase orders, manufacturing orders, and bills of material may be processed by multiple enterprise resource planning (ERP) systems. Hence, providing information in a timely manner throughout the supply chain can be difficult.

Fundamental decisions based on incomplete information may be less than optimal or even incorrect. Additional materials may have to be purchased at last-minute premium rates, incorrect bills of material may be created, erratic inventory levels may lead to stockpiling and production errors, and delivery dates may be missed. These consequences can have a negative effect on the profitability of all participants and place them at competitive risk. To avoid these problems, many companies are buying expensive enterprise integration software.

The objectives of ERP should include more than enterprise-wide interoperability and consistency. They should also include standardization of functional modules, enhanced reliability, and less need for customization. These objectives are not always achieved, however. Many implementations of ERP require a high degree of customization and extensive changes in business processes. Full implementations in large corporations can take years, and sometimes are not completed or are abandoned entirely.

SMEs may want to defer acquiring even MRP systems. These decisions should be based on assessments that include more than initial system and installation costs. The burden of keeping systems operational, plus the costs of periodic updates, may outweigh system benefits to small manufacturers.

There are many degrees of demand planning integration, ranging from manual sharing of information to integrated reporting of demand and electronic forecasting. As improved supply chain integration software and ERP solutions are introduced, the pressure on participants to embrace technological integration will increase. Implementing supply chain management systems is becoming more difficult because of the rapid growth of ERP systems, which are intended for use by single companies. Supply chains include multiple companies and, therefore, multiple (often incompatible) ERP systems. Thus, SMEs in multiple supply chains are being faced with conflicting demands to use expensive systems that may be incompatible with each other. Communications between incompatible systems can usually be accomplished by use of hypertext markup language (HTML), electronic data interchange (EDI), and extensible markup language (XML), but these bridges tend to be awkward, as well as costly to install and operate. Furthermore, XML tends to be field-specific. Substantial progress has been made in the financial and medical fields, but little progress has been made to date in manufacturing. Hence, for the foreseeable future, a practical goal for many small, resource-limited suppliers will be to transmit, receive, and handle information in a reasonably timely, effective, and accurate way, generally without the use of complex ERP software.

e-Business

Internet-based technologies deliver ubiquity (universal and high-speed access), economy (cheap, paperless information and transactions), and utility (platform-independent software and services). The growth in Internet usage and business-to-business e-commerce has been dramatic and cannot be ignored. Buyers are still purchasing the same amount of goods, but an increasing portion of these purchases are being made on line. In 1998, businesses spent $43.1 billion on Internet-based purchases from other corporations (*Wall Street Journal*, July 12, 1999). Forrester Research predicts that by 2003, business-to-business e-commerce will generate $1.3 trillion in revenues, 12 times the estimated consumer market. (Figure 9-2 shows the extent of these worldwide trends.) However, as of mid-1999, only 25 percent of small businesses had Web sites. Concerns about security and underutilization are often cited as reasons for delays in adopting this technology (Grover, 1999).

e-Business technologies can (1) enable direct access to a worldwide, on-line customer base; (2) break down barriers of time and visibility; (3) change the product demand profile; and (4) enable changes in the approach to selling, order taking, and customer service. They can also (1) provide the means for flexible, easy-to-use responses to customer requirements; (2) improve support to product distribution channels, enabling appropriate inventories, improved sales information, and training; (3) provide simplified direct access to supplier products and services; (4) reduce overhead costs for transaction processing; (5) improve access to procurement/financial processes and status; (6) reduce cycle time and cost; and (7) enable the creation of a flexible/agile supply chain with increased synchronization and visibility of inventory.

Although information technologies are transforming many industries, the transition from direct salespeople to e-commerce, for instance, has been slow in coming to SMEs. As of early 1999, more than 90 percent of SMEs with Web sites used them to provide company information, but less than 30 percent used them to sell products and provide on-line customer support. For those who did sell on line, the revenue was a negligible percentage of total revenue (*Wall Street Journal*, August 17, 1999).

Cost has been one reason for the slow change to e-commerce. Managing an on-line operation can create expensive technical problems and

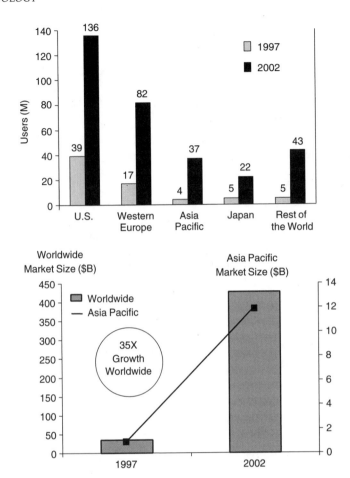

FIGURE 9-2 Internet trends. Source: Murphy-Hoye, 1999.

divert management attention that might be better spent on sales or strategy. The creation of e-commerce sites typically cost more and required greater effort than anticipated. Starting a major new e-commerce Web site in 1998 cost an average of $1 million (*Wall Street Journal*, May 27, 1999). However, SMEs can no longer use cost and lack of computer expertise as excuses for not taking their businesses on line. By late 1999, the cost to set up a small e-commerce Web site had fallen to as low as $100, and the cost for support services per month is about the same. Knowledge of HTML is no longer required. Web site service companies provide templates and instructions in layman's terms, and, for as little as $40 per year, the Web address and a few words about the company can be inserted into hundreds of Internet search engines and directories.

A typical do-it-yourself process for establishing an SME e-business Web site involves the following steps:

- First, build a Web site to display products and take orders electronically. Select a Web site service company. Then choose from among predesigned templates. Drag and drop them onto the site, following simple point and click instructions that make the process almost as easy as sending e-mail.
- Second, contract with a service company to host the Web site. This eliminates the need for acquiring the programming skills and equipment to operate a Web server. The small monthly fee typically includes hosting and basic technical support, as well as the opportunity to create new pages and displays as needed.
- Third, the SME must advertise and promote the site. The site and its products must be listed or cross-referenced on key search engines where on-line buyers can find them. This is a critical step. SMEs must work closely with the service provider to ensure that the proper links are created to channel targeted customers to the Web site. In addition, the SME must go through conventional sales channels to make sure that existing customers are aware of their on-line presence.
- Fourth, because most OEMs will only purchase from prequalified suppliers, SMEs must continue to work closely with customers to get their products approved.

A number of services are designed to help SMEs create and operate Web sites. These services include Sitematic Express (sitematic.com), Homepage Creator (ibm.com/hpc), Bigstep.com (bigstep.com), Virtual Office (netopia.com/software/nvo), Yahoo! Store (store.yahoo.com), WebStore (icat.com), and Internet Store (virtualspin.com).

Recommendation. Small and medium-sized manufacturing enterprises (SMEs) should keep abreast of customer expectations regarding on-line responsiveness and use e-business service providers to assist them in creating and operating low-cost Web sites for displaying products, accepting orders, and answering frequently asked questions. SMEs without an on-line presence may increasingly find themselves at a strong competitive disadvantage.

Supply Chain Integration Software

Supply chain integration software is an expanded, interactive version of MRP and logistics systems that enables companies to collaborate with suppliers and customers, forecast jointly with greater accuracy, shorten product development and introduction cycles, and reduce inventories. These packages, which are often tailored to specific industries and applications, can include tools for electronic data interchange among supply chain partners and modules that control various functions, such as warehousing, purchasing, inventory control, and transportation. Many systems, even from different industries, have elements in common, including (1) a supplier module incorporating multiple layers of suppliers; (2) an operations module consisting of purchasing, inbound logistics, and manufacturing; (3) a customer module, including distribution of goods and services to multiple customer tiers; and (4) a returns channel for optimizing the handling of defective, warranty, trade-in, and obsolete products.

Using these systems to integrate on a function-by-function basis enables the detailed examination and optimization of results by function across the entire chain. Inventory levels, for example, can be given the visibility required to make coordinated decisions rather than forcing participants to hold redundant inventories to buffer lead times. More extensive integration can achieve greater benefits by optimizing on a wider scale and performing more complex trade-offs. For instance, customer satisfaction can be optimized while balancing service levels, asset utilization, and total supply chain costs. More sophisticated packages integrate and optimize processes across elements of the entire chain. Some contain tools to aid in understanding and coordinating interrelated processes in the overall architecture, including process mapping across successive levels. Modeling and simulation of the supply chain can be performed through features that take advantage of ABC and other productivity analysis tools.

Before purchasing supply chain integration software, companies must decide whether (because of unique requirements, the desire to continue present business practices, or the hope of gaining a competitive advantage) to develop custom modules, often at considerable cost and risk. The alternative is to use a standard software package, which, in many cases, requires that users change business processes to fully utilize its capabilities. Many of the standard packages require further development and are far too complex for most SMEs. Therefore, the highest criteria for SMEs considering purchase of software systems for managing their supply chains should be ease of installation and ease of use. When SMEs are faced with OEM demands to integrate with their systems, they should carefully analyze the costs and benefits, as well as the strategic implications, including whether their corporate independence would be jeopardized by extensive integration with a single customer.

Recommendation. Despite significant media coverage of the capabilities of business management systems, small and medium-sized manufacturing enterprises should evaluate, but generally defer, purchasing enterprise resource planning and supply chain integration software until prices come down, these systems are easier to install and use, and the benefits of specific systems have been more thoroughly validated.

Recommendation. Regardless of the level of integration, senior management in small and medium-sized manufacturing enterprises should take the lead in using Internet technologies within their companies. They should closely monitor changes in information technology, invest now in basic capabilities, plan for future investments to support their competitive position, and study how and when to integrate their systems with those of other supply chain participants. Senior management should define data requirements and closely manage the implementation of appropriate data management and electronic communication capabilities.

PRODUCT DESIGN TECHNOLOGIES

The design world is adapting to greatly reduced product cycle times and intensifying time-to-market pressures. Missing a product introduction window in fast-moving industries can completely undermine profitability. CAD, CAM, computer-aided engineering (CAE), design for assembly, design for manufacture, and modeling and simulation systems have significantly reduced the time required for product realization in many industries. Computer-integrated manufacturing (CIM), which enables engineers to take information from CAD systems and use it in a CAM environment, has made it possible to unify all of the basic computer-aided processes involved in manufacturing. With CIM technologies, a single set of product data can be used across a wide spectrum of applications.

Mold designs, welding paths, and computer numerical control (CNC) cutting paths for milling machines and lathes can be generated directly from data derived from solid models created in state-of-the-art CAD applications. With parametric design technologies, the information that shapes the model can be altered easily and quickly. Using these techniques to modify a product design before a commitment is made to manufacture has significantly reduced the time and costs of retooling manufacturing equipment. Virtual manufacturing pulls all of these technologies together into an agile manufacturing enterprise, a virtual factory on a computer that can analyze and pinpoint flaws in the manufacturing process before they occur on the factory floor.

OEMs in integrated supply chains are increasingly asking suppliers to become more involved in all phases of the product realization process. In a similar manner, SMEs should involve their customers and suppliers in the design process. In many cases, involvement in the design process is a new role for SMEs, offering them opportunities to add substantial value to the capabilities they already provide and linking them more closely with the OEM. Early participation by suppliers can enable better designs for manufacturability, provide better opportunities for implementation of advanced materials and processes, and provide sufficient time for the simplification of tooling. Many SMEs have had to expand their design capabilities to participate. For example, GM recently announced that the new Chevrolet Silverado pickup truck was built without traditional clay or wood models. The truck was designed on workstations with engineering software that enabled vehicle, parts, die, and plant designers to work simultaneously on the new product. Suppliers needed enhanced modeling and simulation capabilities to participate (Manufacturing Engineering, 1999).

Modeling and Simulation

Led by the automotive, aerospace, and defense industries, what began 20 years ago as CAD and CAM has expanded into highly complex modeling and simulation capabilities that use interactive design and development tools to reduce the costs and time required for product development and realization. The U.S. Department of Defense is beginning a major initiative called Simulation-Based Acquisition (SBA), the objective of which is complete modeling and simulation of all major weapons systems prior to manufacturing. Simulation-based design, a segment of SBA, uses a digital knowledge environment to represent physical, mechanical, and operational characteristics of these complex systems. In addition to electronically integrating product and process development, prototypes can be tested in virtual environments prior to fabrication. In response to this initiative, prime contractors are increasingly using these advanced technologies and, to the extent that their supply chains are integrated, they will increasingly expect these capabilities from suppliers. When fully implemented, SBA will go far beyond design, modeling all aspects of the product life cycle, from initial concept through manufacturing, sustainment, and even disposal or recycling.

Indeed, in rapidly changing industries, SBA and other techniques, such as concurrent engineering and simultaneous interactive electronic design and development of products, and processes, will offer more and more advantages.

Design System Limitations

Although high-level integrated computer modeling of electrical, mechanical, and manufacturing processes can reduce product realization costs and although "best practice" industry leaders use three-dimensional (3-D) modeling, the most common modeling techniques still involve basic two-dimensional (2-D) CAD designs printed on paper (Integrated Manufacturing Technology Roadmapping Project, 1998). This is partly because of inadequacies in current 3-D modeling systems, including difficulties in translating data between applications. Because only some of the "real information" is in the model, communicating the entire model to others who do not use the same system can be difficult. As of 1999, only a few effective tools were available for collaborative, concurrent use by designers and manufacturing engineers, and most of them were technology-specific or product-specific. Although use of national standards, such as the Initial Graphics Exchange Specification (IGES) and international standards, such as the Standard for the Exchange of Product Model Data (STEP) is increasing, their use is complex and expensive, in some cases requiring the assistance of enterprise network administrators.

The Web is not yet ready for collaborative engineering, although 2-D CAD files can be readily converted into standard Internet (pdf) format, which allows them to be inexpensively posted on the Web or sent via e-mail using the Internet. Until lower cost, user-friendly technology is available, this may be the most practical, cost-effective approach for most SMEs. Real-time transfer of large data objects is difficult to accomplish consistently compared to transfers over privately leased lines. Despite advanced technologies, face-to-face engineering meetings with printouts of 2-D CAD drawings can still be the most cost-effective approach to resolving problems for many SMEs. Although effective standard protocols will eventually be developed for 3-D CAD, SME management must use sound judgement in deciding on the timing and extent of implementation of these technologies.

To date, most OEM attempts to use high-performance modeling and simulation packages have been frustrated by limitations in software capabilities and by slow adoption throughout their supply chains. As integrated supply chains migrate toward higher levels of electronic design, modeling and simulation, and rapid prototyping, SMEs will have to continually reassess their roles. These advanced technologies require substantial investments, and the complexity of leading-edge design tools requires more specialized training and support than many SMEs can afford. The decision can be further complicated because investments in electronic systems to meet the needs of one customer may be of little value in working with another. Furthermore, firewalls must be created so that proprietary information is not transferred between customers. Thus, SMEs should analyze competitors and discuss requirements in depth with customers before making major decisions.

Finding. Advanced electronic systems for product design, modeling, and simulation require further development before they will be practical and cost effective for most small and medium-sized manufacturing enterprises.

Recommendation. Small and medium-sized manufacturing enterprises should carefully analyze the requirements and opportunities for electronic design systems but defer investing in them, if possible, until system capabilities increase and returns justify the investments.

PROCESS AND MANUFACTURING TECHNOLOGIES

The technology needs of an SME extend beyond electronic communications and design capabilities. SMEs must also remain competitive in materials, processing, fabrication equipment, and all of the other technologies used in their businesses. SMEs in fast-moving, high-technology industries must continually reassess their positions and fill capability gaps. In extreme cases, SMEs in high-technology industries may have to essentially reinvent themselves, their products, and their processes as often as several times per decade. They must focus consistently on the latest technologies, whereas in slower changing or mature industries, the appropriate focus is more likely to be on reliability, cost, and delivery. In most cases this does not mean cutting-edge, or "bleeding"-edge, capabilities; others can pay the price of being the first to debug a new technology. Accepting the risk of trying to be first only makes sense in industries that greatly reward early adopters of new technology. However, implementing carefully selected advanced technical capabilities may be one of the few ways of enhancing profitability in markets with increasing price pressures. For example, a company in the precision forging business may rely on expertise in advanced precision forging techniques to differentiate itself from its competitors.

Competitive advantage can be gained through a combination of investments, experience, research, experimentation, and operator training. The development of innovative process and manufacturing technologies may offer opportunities for added value, although the costs of such development may be beyond the reach of financially limited SMEs.

Agility, flexibility, and responsiveness are becoming increasingly important in the current climate of rapidly changing customer and supplier needs. Flow manufacturing is a new approach that offers improved speed, response, and flexibility throughout the production, procurement, and order fulfillment process. The basic premise of flow manufacturing is the pull of materials through production and the supply chain based on actual customer demand, rather than the push of materials based on a preset schedule (Blanchard, 1999). Flow manufacturing encompasses many Japanese lean manufacturing techniques, such as reduced cycle times, reduced inventory, mixed-mode manufacturing, and line balancing. The strategy uses a planning horizon of several hours or several days rather than the 12-week horizon of traditional production planning.

Recommendation. Despite the increasing importance and glamour of Internet-based technologies, small and medium-sized manufacturing enterprises should not ignore up-to-date manufacturing and process technologies. They remain essential for success.

SOURCES OF TECHNOLOGIES

Technologies for supply chain participation are available from a variety of sources:

- Many universities are eager to license technologies developed in their laboratories. They are also willing to establish cooperative research and development agreements to develop technologies of interest to their academic staff. Thus, at reasonable cost, SMEs can have access to the same highly skilled people as large OEMs.
- Government laboratories in the National Aeronautics and Space Administration, the U.S. Department of Defense, and the U.S. Department of Energy are under incentive to license, at reasonable rates, non-classified technologies developed with public funds.
- Modern communications have made it easier to learn about technologies developed in foreign countries. Some of these technologies are excellent and can be licensed. Representatives of the former Soviet Union, for example, are eager to sell or license technologies, especially for Western currencies.
- Skilled employees and recent graduates can contribute greatly to the development of state-of-the-art technologies. Highly skilled

immigrants from economically depressed parts of the world can be hired at competitive rates. International centers, such as Washington, D.C., New York City, and San Francisco, are especially attractive to skilled immigrants, and it may be worthwhile to locate an operation there to access their skills.

- Acquiring a small start-up company with a key proprietary technology may be less expensive than risking time and money to develop a competing technology.
- Courses, seminars, conferences, and technical publications can alert and educate employees to state-of-the-art technologies and opportunities for innovation.
- SMEs in possession of key proprietary technologies have potentially valuable bargaining chips if these technologies are needed by OEMs. The benefits of such a business relationship might include technology cross-licensing, engineering assistance, subsidized manufacturing facilities, and low-cost installation of compatible MRP and electronic design systems.

FINANCIAL ISSUES

Successful participation in many integrated supply chains is becoming increasingly difficult for SMEs unless they have extensive financial resources. Keeping up with new technologies, the increasing demands of supply chain integration, the increasing risks of product liability, and the reserves necessary to respond to rapid changes in the business environment all require strong financial reserves. Substantial investments in capital equipment and training may be required to remain competitive.

OEMs and higher tier suppliers are increasingly demanding that SMEs make specific investments, many of which are not one-time investments. Systems can become technologically obsolete within a few years, and customers may insist on upgrades even sooner. Although 15-year-old production equipment is common, seven-year-old computers are obsolete. SMEs must find ways to meet these demands. Moreover, as OEMs consider establishing long-term relationships with their suppliers, they are evaluating their financial underpinnings and track records more closely. Thus, SMEs must demonstrate a track record of financial stability and implement financial systems to control costs and inventories.

Although financial requirements were not high on the list of SME concerns in the survey conducted for this study, preventing and filling financial gaps is a central issue. To reduce the need for outside funding, SMEs must focus their activities carefully, maximize their cash flow, invest their resources wisely, and take advantage of the techniques of supply chain integration to reduce the amount of capital tied up in excess inventories and excess manufacturing capacities.

Funding Sources

SMEs must have financial resources in place in advance of need. Establishing sound investor and banking relationships in advance is essential to long-term financial stability, and like all partnerships, these relationships require ongoing attention. Venture capitalists are a potential source of funds, although they are unlikely to fund an SME unless it exhibits an unusual potential for rapid, highly profitable growth.

Large OEMs and supply chain partners sometimes provide resources and "co-investments" for key SME participants, thereby enabling the SME to become a better supplier, strengthening the partnership, and ultimately benefiting the OEM. This strategy has been pursued successfully by Japanese OEMs, such as Canon and Toshiba. OEM investments can take the form of advance payments for products, equity investments, loans, subsidized product and process development, technology transfer, compatible electronic design and MRP systems, training, and access to experienced people. In the past, Ford, for example, often bought conventional machinery for supply chain members if they needed it to produce for Ford. However, as of early 1999, Ford purchases machinery for suppliers only if the machinery is unique for Ford requirements.

Few SMEs have used supply chain management techniques to integrate their own supply chains. These techniques can reduce the need for investing in redundant inventories and excess manufacturing capacities, thereby freeing cash for other investments.

Recommendation. As supply chain integration requirements and the need for new technologies increase the financial requirements imposed on small and medium-sized manufacturing enterprises, they should integrate their own supply chains to reduce redundant inventories and excess manufacturing capacities, thereby freeing cash for other investments.

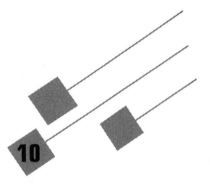

Globalization and Proximity

GLOBALIZATION

Sourcing has become increasingly independent of location. The days of a supplier matching prices with a few local rivals are disappearing, and many SMEs now compete with a multitude of suppliers worldwide. Increased global demand for customized products flows throughout integrated chains, requiring flexibility, responsiveness, and the ability to customize. As OEMs compete on a global basis, their supply chains must support them with global capabilities. This support can take the form of overseas factories, warehouses, and distribution centers; the production of parts with metric dimensions; certification to international standards; learning to work effectively with other cultures; the use of communications systems that transcend global distances; and the use of modern transportation methods. SMEs may also have to expand their own supply chains internationally, procuring goods and services wherever they can be most competitively obtained. For example, Fusion Lighting, an SME in Rockville, Maryland, procures power supplies from Sweden, die castings from Taiwan, magnetrons from Japan, and reflectors from Germany and then ships finished lighting products to OEM customers in the United States, Japan, and Sweden. The custom die castings from Taiwan, although equal in cost, delivery, and quality to those made in the United States, are produced using tooling that cost one-half the going price in the United States.

SMEs seeking to expand into foreign markets should carefully analyze the full costs, risks, and benefits of such a strategy. Political requirements regarding local content, sales ventures, and repatriation of funds, as well as the risks associated with fluctuating currency exchange rates, can have significant effects on competitiveness and profitability.

PROXIMITY

One of the key differences between traditional supply chains and highly integrated supply chains is the degree of proximity between members and the resulting differences in the efficiency of joint operations. Geographic and cultural proximity traditionally provided business advantages for SMEs, many of whom served only local customers and had to compete only against other local suppliers. Globalization, electronic communications, and modern shipping capabilities now enable suppliers from all over the world to compete for local business. Large suppliers can typically afford proximity capabilities that SMEs cannot, including plant sites near their customers and skills in dealing with different cultures. To remain competitive, SMEs may have to improve their organizational, cultural, and geographic proximity to serve an increasingly widespread customer base.

Organizational proximity can take several forms, including membership in joint project teams or the placement of employees in one another's facilities. Cultural proximity, which typically evolves over time, can be achieved through the adoption of common business practices, jargon, ethical standards, and language. Cultural proximity is especially important for doing business with customers from different countries and cultures. The dividends of cultural proximity can include repeat business, loyalty, and assistance in problem solving during times of crisis.

Geographic proximity may involve locating supplier facilities adjacent to OEM operations. For example, long-term relations between beverage producers and container manufacturers led container suppliers to locate their fabrication plants adjacent to breweries. Cans are drawn, finished, and moved on conveyors through a common wall into the brewery where they are filled, sealed, packed, and shipped, all without human contact.

Proximity in an international supply chain can require investments in metric dimensioning, compliance with international standards, and participation in international trade fairs, such as the one held annually in Hanover, Germany. Partnerships with foreign companies can be used by SMEs to obtain cost-effective access to foreign markets and sources of supply. Effective international participation requires knowledge of the ways of doing business in other countries and cultures. International consultants can sometimes fill these gaps, but management must be appropriately trained, especially for face-to-face communications and negotiations. Training programs, such as the Massachusetts Institute of

Technology (MIT) Japan Effectiveness Training course, can be helpful in this regard.

ELECTRONIC ALTERNATIVES

The Internet and the Web are great equalizers in proximity and global presence, providing equal opportunities for suppliers large and small, foreign and domestic. A presence on the Web is becoming increasingly important for many SMEs because it can deliver the corporate message to a global audience and increase business opportunities. It can also increase risk, however. Advertising the latest capabilities, product features, and prices informs both customers and competitors. Requests for product samples may come from potential customers or, either directly or through third parties, from low-cost competitors in countries with scant regard for intellectual property rights.

Because most SMEs are not in a position to locate plants near each of their customers, they must find other ways to deal with the issues of proximity. Some of the problems of distance can be overcome by means of electronic systems, such as e-mail, electronic data exchange, and video conferencing, which can provide some of the benefits of proximity at lower cost. Supply chains in the computer industry, for instance, are typically geographically scattered and have little organizational or cultural proximity, but they do have close electronic proximity. However, no matter how advanced the communications system, true geographic proximity always provides an advantage because there is no substitute for face-to-face contact. "Take the look in the customer's eye when you tell him a new price," says Thomas W. Malone of the MIT Sloan School of Management. "That's very useful information" (*Washington Post*, March 6, 1999).

Recommendation. Because Internet technologies and modern transportation capabilities enable suppliers from low-wage areas to compete effectively with U.S. small and medium-sized manufacturing enterprises (SMEs), pressure on SMEs to improve their cultural, organizational, and geographic proximity to their customers and suppliers has increased. Even SMEs with limited resources can respond to some of these challenges at low cost through increased cultural education and use of the Internet and Worldwide Web.

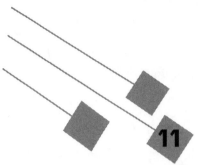

Requirements and Characteristics of Successful Small and Medium-Sized Manufacturers

EVOLVING REQUIREMENTS

Because the business environment and product requirements keep changing, the requirements imposed on suppliers and the processes of supply chain building and integration are also changing. Young industries typically focus on providing new, innovative products, and their supply chains must be flexible enough to respond to rapid changes in technology and market demand. As the industry matures, the primary basis of competitive advantage shifts to other capabilities, such as marketing and distribution. As the industry continues to mature, the emphasis typically evolves toward cost saving, a limited product array, high levels of customer service, and improved manufacturing practices. Thus, SMEs must evolve the focus of their enterprises in a manner compatible with the phases of the industries they serve.

The roles of technology evolve in a similar manner. For instance, during the first 70 years of automobile manufacturing, electrical systems played a subordinate role. Design and fabrication of the metal structure constituted the core technology, and electrical devices were considered secondary. Today, the cost of automotive electronics exceeds the cost of the steel structure. These changes have led to parallel changes in the required competencies and associated business opportunities for SMEs. Thus, suppliers must be constantly aware of changing requirements and opportunities.

The business environment in most industries is evolving far more rapidly than ever before, and competitors elsewhere in the world are combining low-cost labor with improving education, communications, and transportation capabilities to compete for roles in OEM supply chains. For instance, in 1997, a New York apparel manufacturer advertised a spring fashion line on his Web site. Within days he had five orders from Beijing. The manufacturer at first savored the irony that China, a growing source of garments sold in the United States, was buying from New York. Soon, however, he began to suspect that his garments had been bought to be copied. Suspiciously similar dresses with "Made in China" labels flooded the market within a month (Kanter, 1998). Thus, U.S. SMEs can lose their competitive position virtually overnight.

Although changes like these have not yet affected all industries or all levels of supply chains, competing in fast-moving environments will require new strategies. Departments and functions within the corporation and throughout the supply chain must work simultaneously, rather than sequentially, on projects. The feedback loop between the marketplace, designers, manufacturers, and the sales force must be shortened so that products can be developed and transitioned to manufacturing faster and introduced simultaneously into markets around the world. Electronic data interchange capabilities may be required to enable key supply chain functions to respond simultaneously to changes in customer demand. Flexible or agile manufacturing may be needed for rapid responses to changes in demand, timely resolution of quality problems, and fast implementation of new product features. SMEs will have to assess their future roles carefully and may have to reposition themselves rapidly to remain competitive. In some cases they will have to make fundamental decisions about whether to (1) accept the risks associated with investing and transforming the company (see the Case Study below); (2) sell the business while it still has some value; or (3) operate the business essentially as is until it fails.

Recommendation. Small and medium-sized manufacturing enterprises must reassess their competitive positions and those of their supply chains on a regular basis and position themselves to respond rapidly to changing conditions.

Case Study: Astronics Corporation's Repositioning Strategy

Astronics Corporation, an SME headquartered in Buffalo, New York, operates in two distinct business segments: specialty packaging and aerospace electronics. In the 1980s, the packaging segment served as a regional supplier of custom folding boxes. The electronics segment has been selling primarily to higher tier aerospace and defense suppliers and to OEMs. Astronics' revenues in 1986 totaled $17 million.

Physical proximity had been a valuable differentiator for the folding box supplier in terms of shipping costs and face-to-face contact. As this commodity business became increasingly competitive, Astronics sought to differentiate itself and add value by investing in capabilities that enabled it to pursue markets in short-run specialty packaging and high value-added consumer products. The company invested heavily in innovative production technologies, including lasers and water-jet cutting. It also converted to computerized plate and die-making equipment, which reduced tooling cycle times by more than 80 percent. These investments, plus the shift in target markets away from low-value commodities, led to substantial increases in revenues and earnings. To maintain this growth, the company plans to consolidate purchasing, invest in new production capabilities, offer a wider range of products, and enter into innovative supply chain alliances with suppliers and customers.

The markets served by the aerospace electronics segment are characterized by rapidly changing customer demand (contracts) and by technology changes. To remain competitive, Astronics selects its product development contracts carefully and keeps its technological capabilities up to date. Astronics has carefully invested in the development of this segment by means of acquisitions and internal development.

As a result of these efforts in both divisions, 1998 corporate revenues totaled $46 million, an increase of more than 250 percent in 12 years, and profits have increased steadily. Thus, by analyzing the changing business environment, repositioning itself to meet evolving challenges and opportunities, adding value to its products, and judiciously investing in technologies, this SME supplier has been able to grow and prosper (Astronics Corporation, 1995, 1998).

CHARACTERISTICS OF SUCCESSFUL SMALL AND MEDIUM-SIZED MANUFACTURING ENTERPRISES

The committee interviewed executives of several successful SMEs in an attempt to identify common characteristics that are essential for success. The SMEs, which ranged in size from $2 million to $50 million in annual sales and had 30 to 450 employees, were selected because they have viable businesses, strong customer bases, and positive out-year projections for sales and profits. Based on these interviews and the experience of the committee, successful SMEs have several characteristics in common, which are discussed below.

Evaluation of Customers

Successful SMEs are good at selecting customers. To a considerable degree, the interviewed SME executives had selected the customers with whom they do business and had turned away customers that treated them as adversaries. At first glance, this might seem like an extreme approach for SMEs with only a few customers, but the development of a diversified base of loyal, trusting customers can reduce business risk during hard economic times. The SME executives felt that careful evaluation and selection of customers was one of the most important criteria for success.

Because participation in integrated supply chains is increasingly a long-term commitment, customers should be assessed carefully to determine their reliability and the appropriateness and long-term viability of becoming integrated into their chains. Each SME should consider the following questions and create detailed customer evaluation criteria specific to its business needs:

- Does the customer offer multiyear contracts? Spot buys may be good for short-term revenue, but the benefits of an integrated supply chain can be maximized only when both the customer and SME make long-term commitments.
- Does the customer practice the principles of successful supply chain integration? How willing is the customer to work with SMEs to improve supply chain processes? Is the customer agreeable to equitable sharing of benefits and burdens?
- Are the customer's employees evaluated, in part, by how well they interact with suppliers? The SME should determine if the customer has an internal environment conducive to successful integration. In an integrated chain, the SME will have a close business relationship with the customer's employees, and the way they are evaluated will affect how they interact with suppliers. SMEs should also note how the customer treats others, both under normal conditions and under conditions of stress. Successful integration requires that partners be willing and able to work effectively together under all conditions.
- Is important business information shared openly between customer and supplier? Successful integration depends on sharing information. SMEs should assess the types of information that are shared and the means by which they will flow. Does the customer demand more information than is needed for successful integration? Access to customer schedules and other aspects of their MRP systems can enable an SME to position itself to respond to customer needs. If the available information does not meet the needs of the SME or is not available in a timely and useful form, the likelihood of successful integration should be seriously questioned.

- What forms of communication does the customer require? Because many large OEMs are requiring more data and compatible electronic connectivity with suppliers, SMEs should evaluate the costs of gathering the required data and transmitting it by means of compatible systems. They must determine if these investments are compatible with their long-range plans.
- What access does the supplier have to the customer's product plans and technology strategies? What are the OEM's expectations with regard to supplier participation in product design and development? To be truly integrated and effective, SMEs must understand the customer's product plans and participate at an appropriate level in product design.
- To what extent does the customer demand and recognize quality certifications, standards, and processes? Customer requirements should be appropriate for the products being purchased, and the supplier's quality capabilities should be compatible with the customer's needs. Supplier conformance with customer requirements should be recognized and supported by customer actions, such as fewer site audits.
- Can the customer be profitably served? SMEs should assess requirements for additional investment and the extent to which the customer's needs will be compatible with the needs of its other customers. Assessments of the profitability of serving each customer require the capability to isolate costs by customer or supply chain. Thus, the committee suggests that SMEs consider implementing ABC (activity-based costing). By adopting net present value and return on investment approaches, an SME can quantify and compare the costs of serving various customers and estimate the revenues and profits that will accrue from the relationships.

Recommendation. Small and medium-sized manufacturing enterprises should develop and implement customer evaluation and selection criteria, including the following:

- opportunities for long-term contracts
- approach to supply chain integration
- extent of information sharing
- required forms of communication
- access to product and technology plans
- expectations regarding participation in product design
- recognition of supplier quality certification
- whether the customer can be served profitably

Reaction to Salient Events

One key to success identified by all of the SMEs who were interviewed was the ability to recognize and react appropriately to salient events (major events that can define success or failure). For example, one SME, the Texas Nameplate Company, Inc., reported an initial negative reaction when the company was forced to adopt statistical process controls. However, when the benefits of successful implementation were recognized, the SME elected, on its own initiative, to adopt total quality management (TQM) and ISO 9000. It later went on to win the prestigious Malcolm Baldrige National Quality Award. Thus, customer-imposed requirements that may seem onerous in the short term might prove to be beneficial in the long term.

Strategic Alliances and Partnerships

Successful SMEs create, implement, and maintain strategic plans, which include partnerships and strategic alliances that enable them to meet customer needs better. These alliances should be formalized in documents that define roles and responsibilities. Strategic alliances and partnerships increase the likelihood that the customer will use the SME for future business.

Catering to the Customer

The SME must cater to the customer's needs, which increasingly include supply chain integration. This means, first and foremost, providing low-cost, high-quality products, effective service, and on-time delivery. SMEs should know their customers intimately at all levels in the organization and cater to specific business needs at each level.

Focus on Quality

Successful integration requires that supplier quality levels be appropriate to meet the customer's needs and that the attitude toward high quality be instilled in all employees.

Treatment of Employees

Employees, the interfaces between customers and suppliers, must be educated and trained to meet the skill levels required by increased supply chain integration. SMEs should use incentives to enhance the performance and reinforce the training of employees.

Selection and Monitoring of Metrics

SMEs should identify which aspects of their business should be measured, determine meaningful metrics for measuring progress, and use the information to improve performance. Collecting data is expensive, however, so SMEs should only collect and analyze data that will help to improve business operations.

Documentation of Business and Manufacturing Processes

Documentation is a foundation for quality. The ability to repeat standard, well documented processes without deviations is crucial for long-term success.

Use of the Internet

The Internet can be an inexpensive way to meet communications requirements. It also provides opportunities for education, new sales channels, and lower cost procurement.

Sharing of Information

SMEs should develop the capability to provide accurate, robust data to supply chain partners. Supply chain integration/management software should be chosen carefully. Most of the software is much too complex, expensive, and time-consuming for SMEs.

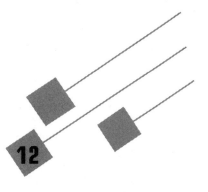

Assistance for Small and Medium-Sized Manufacturing Enterprises

How can SMEs systematically identify specific constraints and performance gaps that inhibit their competitiveness? Where can they turn for assistance? One approach is to develop close relationships with other members of the supply chain in hopes of benefiting from their experience. Each company, however, has its own agenda, and diverting resources to teach other companies is not usually advantageous. Conferences, trade shows, consultants, trade journals, and industry groups can be helpful. Among the best resources are manufacturing extension centers and technical resource providers (MEC/TRPs), which are chartered specifically to provide advice and counsel to SMEs. MEC/TRPs are typically financed by combinations of public and private funds and can be found in virtually every city and region of the United States. Local politicians can usually assist companies in finding these organizations.

MANUFACTURING EXTENSION CENTERS AND TECHNICAL RESOURCE PROVIDERS

Manufacturing Extension Partnership

The Manufacturing Extension Partnership (MEP), a sponsor of this study, is a nationwide network of more than 70 not-for-profit centers, located in all 50 states and Puerto Rico. Linked together through NIST, they provide access to more than 2,000 manufacturing and business specialists who have hands-on experience. Each center has its own identity, with names like the Florida Suncoast Manufacturing Technology

Center, the Northeast Pennsylvania Industrial Resource Center, and the Great Lakes Manufacturing Technology Center.[1]

Each center has the ability to assess a wide variety of problems, provide technical and business solutions, help SMEs create successful partnerships, and provide seminars and training programs. Combining local expertise with national resources, they have assisted more than 62,000 SMEs with problems, including the following:

- locating resources or technologies
- determining causes of product defects
- modifying plant layouts
- establishing employee training programs
- increasing sales and improving market share
- reducing costs
- implementing new technologies or processes
- managing change
- selecting and implementing business and quality management systems
- adopting information technology to reduce time to market
- identifying and exploiting manufacturing niches
- conducting energy audits and reducing energy costs

The MEP network has positioned an organization called Supply America Corporation to act as a single contact point for the MEP in supply chain integration. Supply America focuses MEP resources on delivering value-added supply chain products and services to OEMs and their suppliers to improve overall supply chain performance.

Robert C. Byrd Institute

The Robert C. Byrd Institute (RCBI), another sponsor of this study, is a national technical service provider. The institute has four sites in West Virginia and serves portions of Pennsylvania, Maryland, Ohio, Kentucky, and Virginia. Sponsored by the Defense Advanced Research Projects Agency, RCBI was established to develop a quality, just-in-time manufacturing supplier base for the U.S. Department of Defense and its prime contractors through "teaching factories," computer integration, and workforce development. RCBI addresses challenges faced by SMEs such as:

---

[1] Locations can be obtained by calling 1-800-637-4634 or by visiting the MEP Web site at *http://www.mepcenters.nist.gov.*

- keeping abreast of changing technologies, production techniques, and business management practices
- difficulty obtaining high-quality, unbiased information
- isolation from other manufacturers, including limited interaction and networking
- disproportionate burdens created by the regulatory environment
- difficulty obtaining operating capital and investment funds
- lack of trained workers, such as certified machinists
- implementing quality certification programs and new technologies

RCBI's goal is to help SMEs (1) to increase productivity through new technologies, (2) to improve competitiveness through workforce development, and (3) to maximize return on investment by integrating technologies. RCBI has developed innovative ways to improve the competitiveness of SMEs at reasonable cost. For instance, they assist SMEs in achieving quality certification for specific task areas designated by an OEM. Success at the task area level often increases an SME's willingness to commit resources for an entire ISO 9000 certification. RCBI's operating units work directly with SMEs to help them identify and fill gaps in their capabilities. The Technical Training/Workforce Development Unit, for example, provides a variety of educational initiatives, including customized training, seminars, workshops, and technical assistance in the following areas:

- quality programs, including ISO 9000, QS 9000, AS 9000, quality system documentation, and SPC
- management programs, including leadership development, team building, and task analysis
- job training manuals
- design and manufacturing programs including CAD/CAM training, CNC instruction, programmable logic controller training, welding certification, and safety programs
- nationally certified machinist training

The Technical Services Group offers SMEs an opportunity to learn about flexible manufacturing systems, including operator training, while actually producing a product. This "teaching factory of the future" allows SMEs to take advantage of new technologies, increase sales, and improve market share. The Systems Integration Group is a consulting and project management/implementation resource for computers, networking, and telecommunications projects. The 21st Century Manufacturing Network brings the areas of electronic commerce, electronic data interchange, technical education, and computer and network systems integration together to provide manufacturers with a central, "virtual" location for business assistance resources. This network of more than 150 regional manufacturers is helping SMEs to create an infrastructure for conducting business on line.

Northeast Tier Ben Franklin Technology Center

The Northeast Tier Ben Franklin Technology Center, located in northeastern Pennsylvania, is an example of a small regional center. One of four regional centers operating under the Commonwealth of Pennsylvania's Ben Franklin Partnership, its mission is to lead northeastern Pennsylvania to a better economic future by assisting clients in creating innovative solutions that integrate people, technology, and systems for novel competitive advantage. Funding comes primarily from the state and from public and private matching funds. Working closely with Lehigh University, the center's goal is to help companies add value to the products and services produced in the region so that they can compete more effectively with producers from low-wage countries.

Virginia's Manufacturing Innovation Center

Virginia's Manufacturing Innovation Center, sponsored by James Madison University and the Center for Innovative Technology, is an example of a state organization focused on helping SMEs. Its mission is to enhance the competitiveness of Virginia's SMEs through the development of a well trained workforce and by providing access to advanced computing technology and modern management practices. The center will house laboratory and training facilities (*www.isat.jmu.edu/vmic*) including:

- the Integrated Learning Factory, a modern production facility to develop and demonstrate computer-based automation and integration technologies
- the Biomanufacturing Training Facility to develop skills required for the design and management of biopharmaceutical manufacturing facilities
- the Microfabrication Laboratory, a clean-room facility used to demonstrate the fabrication of integrated microelectronic devices, sensors, and microelectromechanical systems
- the Manufacturing Management Laboratory, which provides hands-on learning experience in the dynamic and integrative nature of managing production operations

Strengthening Manufacturing Extension Centers and
Technical Resource Providers

The committee found that not all MEC/TRPs are fully capable of helping SMEs compete successfully in a rapidly changing integrated supply chain environment, and not all of them are consistently proficient in providing guidance to SMEs seeking to integrate their own supply chains. Specifically, MEC/TRPs must develop a standard set of supply chain best practices for SMEs and implement appropriate support programs at all of their centers. Uniform, high-quality programs are essential because supply chain integration typically involves multiple companies in scattered locations. Therefore, inconsistent local programs and levels of support can make integration efforts difficult. Sufficient funding is essential for MEC/TRPs to carry out this important new mission without detracting from their other operations in support of SMEs.

SMEs generally cannot afford the same high-priced guidance provided by consultants to OEMs. As the survival of SMEs is being increasingly imperiled by converging trends in supply chain integration, technology, and logistics, which are resulting in dramatic increases in low cost, global competition and substantial demands for investment, SMEs have unprecedented needs for state-of-the-art guidance at an affordable cost. Thus, MEC/TRPs should be provided with sufficient public and private funding so that they can focus their efforts on providing critically needed services rather than on fund-raising. SMEs, in turn, should rely more heavily on MEC/TRPs to guide them through their increasingly complex business environments.

SMALL BUSINESS SET-ASIDES

Considering the imposing requirements of some integrated supply chains, it might seem impossible for SMEs to participate, even in the lower tiers. However, Public Law 100-656 requires that OEMs receiving government contracts in excess of $500,000 give preference to, or put a good faith effort toward, setting aside as much as 20 percent of their subcontracts (or the priced bill of materials) for small, disadvantaged, and minority-owned businesses. Of this 20 percent, 5 percent is allocated for small disadvantaged businesses and 5 percent for businesses owned by women. Numerous OEMs, despite extensive efforts, never fully meet these quotas, often because of a lack of qualified candidates. Many OEMs have taken steps to assist SMEs by making provisions for their participation on procurement review boards, sponsoring small business supplier conferences, and establishing mentor protégé programs to assist SMEs in process development and the development of program plans and budgets.

McDonnell Douglas (now part of The Boeing Company), for example, offers the following courses at no fee to its suppliers:

- Statistical Process Control
- Quality Function Deployment/The Taguchi Approach
- Design for Manufacturability
- Design for Assembly
- Benchmarking
- Preferred Supplier Certification
- Effective Presentation Seminar
- Developing Team Performance
- Design, Manufacturing, and Producibility Simulation

Even though small business set-asides provide opportunities for participation, they do not ensure success for SMEs. Success still requires good performance.

Recommendation. Small and medium-sized manufacturing enterprises (SMEs) should avail themselves of the opportunities provided by government small/disadvantaged business programs to accrue financial resources, develop skills and capabilities, acquire compatible systems, and build trusting relationships so that when they are no longer eligible for special consideration, they can stand on their own as fully competitive and integrated members of supply chains.

OTHER RESOURCES

Other resources that can assist SMEs in filling educational gaps include academic journals, such as the *Journal of Business Logistics, International Journal of Logistics Management, International Journal of Physical Distribution and Logistics Management, Transportation Journal,* and *Supply Chain Management Journal,* and supply chain periodicals, such as *Supply Chain Management Review, Inbound Logistics, Inventory Reduction Report,* and *Global Sites Logistics.*

A variety of educational programs are also available. The Center for Advanced Purchasing Studies and the National Association of Purchasing Management offer a variety of seminars that can be useful for SMEs. Colleges and universities offer seminars, courses, and programs addressing issues associated with supply chain participation, integration, and optimization. SMEs must be selective, however, because some of them are expensive, and some are geared for sophisticated OEMs with massive computer and analytical capabilities.

ASSESSMENTS OF COMPETITIVENESS

A variety of tools can be used by MEC/TRPs to assist SMEs in assessing their competitiveness and identifying gaps in their capabilities. Business opportunities within a supply chain can be evaluated by using the same analytical tools used to evaluate other business opportunities. However, these opportunities should also be analyzed within the context of optimizing the supply chain as a whole.

Mapping

Few SME managers have identified or understand process requirements and capabilities one supplier tier away, much less two or three. Thus, MEC/TRPs should help SMEs map critical segments of the supply chain in terms of organizations, capabilities, and functions, paying special attention to critical and sole-source capabilities. Ideally, these maps should extend to every key capability and function required to design, manufacture, distribute, sell, and support the product line. Specialized maps of evolving technologies, manufacturing capacities, and other strategic functions can be helpful for planning, integration, and problem identification.

Mapping should begin with the identification of key members, functions, and processes of the "neighboring" tiers of the supply chain. Attempts to map, integrate, or manage all processes and functions will generally cause the mapping process to become extremely complex. Thus, at this stage the functions and processes that are most deserving of management attention and corporate resources should be identified and prioritized. SMEs may, for instance, wish to focus only on operational and/or managerial activities that produce specific outputs or add value to the product and disregard support activities. Alternatively, SMEs may wish to focus on non-value-added activities with a goal of ultimately reducing or eliminating them. It is generally worthwhile to include the basic capabilities required of all participants (e.g., quality, cost, service, delivery, basic communications capabilities, fundamental technologies, and financial viability), as well as requirements unique to the specific supply chain (e.g., rapid prototyping, design and development capabilities, enhanced communications capabilities, and unique process requirements). Participants should be evaluated for each required capability and recommendations should be developed to provide specific guidance and priorities to assist participants in improving or acquiring the necessary capabilities.

Over time, the knowledge obtained while assisting SMEs with the mapping process could enable MEC/TRPs and their representatives to become increasingly effective in helping a wide range of SMEs.

MEC/TRPs should collect this knowledge in a readily accessible database, identifying, cataloguing, and updating successful approaches for improving SME performance (see Appendix B for additional information on mapping techniques).

Recommendation. Small and medium-sized manufacturing enterprises should use mapping techniques to identify supply chain requirements systematically, compare them with their own capabilities, and rigorously assess their own gaps and constraints. They should use these same techniques as a means of assessing and strengthening their supply chain partners.

Recommendation. Manufacturing extension centers and technical resource providers should develop and implement formal, rigorous programs for (1) mapping supply chain requirements against the capabilities of individual small and medium-sized manufacturing enterprises (SMEs) to provide them with effective, specific guidance; (2) gathering data to assess their own training programs; and (3) training SMEs in the use of these techniques.

Competitiveness Review

NIST and the MEP have developed a holistic assessment tool based on the Malcolm Baldrige National Quality Award and ISO 9000 called the Competitiveness Review. The review, which covers delivery, cost, quality, management, technology, safety, environment, and other business parameters, can provide fundamental guidance to SMEs in assessing their capabilities and identifying opportunities for improvement. The review emphasizes metrics, corrective actions, employee involvement, and goal setting.

Several factors should be taken into consideration when using the Competitiveness Review. First, because management is a complicated, multifaceted endeavor, it is difficult to isolate and measure the effects of specific management practices. Second, technology pervades all aspects of business. Therefore, measures of technological success must be based on the results of technology implementation and utilization (e.g., whether a new technology leads to improved business performance). This requires a holistic approach that compares the benefits (e.g., increased sales and profits) with the costs of acquisition and implementation. Third, the subject of innovation is included in the section on technology. However, because innovation may or may not involve technology, a broader view of innovation should be considered. Innovation is essential for sustaining a competitive advantage in every aspect of business. The committee chose to include innovation in discussions of management because it is critical that management create a corporate culture broadly conducive to innovation. Fourth, the Competitiveness Review facilitates measuring the current status of an enterprise and identifies targets for improvement for each of the variables. However, it does not identify constraints or help resource-limited SMEs decide which factors are most critical.

Constraints

The identification of internal and external corporate constraints is an important adjunct to the assessment of capability gaps. Constraints can take the form of limited capital resources, lack of employee know-how, conflicting customer demands, inadequate operating policies, resistance to change, poor leadership capabilities, and so forth. Because constraints can be limitations or barriers to improvement, identifying and dealing with the most significant constraints is as important as filling the highest priority capability and performance gaps.

The Goldratt Theory of Constraints attempts to identify the root causes of undesired business effects by evaluating the constraints (both operational and policy) that stifle innovation (Goldratt, 1990). The Goldratt Theory can be used as an adjunct to the Competitiveness Review to help SMEs describe, in operational terms, their goals and how they plan to achieve them. Goldratt used this method, for instance, to analyze a major supply chain for the Cadillac Motor Division of GM, and Cadillac is using the results for competitive advantage.

Thus, the challenge for MEC/TRPs is to help each SME understand the implications of supply chain requirements, set appropriate goals or targets, identify the most significant gaps and constraints impeding its business performance, and suggest ways to overcome them. Capability mapping can be helpful for identifying and prioritizing specific capability gaps. Closing the gaps involves (1) eliminating or circumventing constraints, (2) obtaining appropriate capabilities for the evolving business environment, and (3) using those capabilities effectively.

Recommendation. Small and medium-sized manufacturing enterprises should seek out local manufacturing extension centers and technical resource providers for assistance in understanding the supply chain integration process, identifying constraints and capability gaps, laying out a road map for improving performance, and implementing the road map.

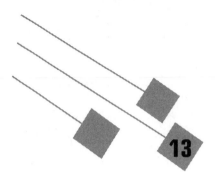

Conclusions

First, what's a convergence? It's the coming together of people or ideas in ways that didn't happen before. And the challenging thing about this process is that when it happens, it always changes the rules of math. It causes one and one to make three. That is, the result of the convergence is greater than the sum of its parts. Especially when it involves the development of new technologies (Burke, 1999).

The specific recommendations in this report are the true conclusions of this study. Nevertheless, some salient conclusions pervade the report as a whole, and it may be useful to make them explicit here. Many of the recommendations are not new to SMEs. However, (1) the demands of supply chain integration, (2) reductions in the number of suppliers to OEMs, (3) changing product design technologies, and (4) the development of the Internet, e-commerce, and modern logistics methods are converging on SMEs (and the MEC/TRPs that support them) at an unprecedented pace. This convergence of multiple trends, coupled with unprecedented rates of change within these trends, is jeopardizing the competitiveness of many U.S. SMEs.

In the past, SMEs that failed to respond to changing trends gradually became less competitive. Today, SMEs may find, almost overnight, that they have been surpassed by global competitors and have lost critical customers. With limited resources and the increasing speed of events, they may be unable to respond. Thus, SMEs must be alert to rapidly changing conditions and respond while they still have a business, some resources, and the time to do so. Many SMEs will require increased assistance from MEC/TRPs to respond effectively.

From a different perspective, this convergence has created unprecedented opportunities for suppliers from low-cost areas of the world. With reduced trade barriers, access to technologies and modern management methods, increased training in English, better educated and more skilled workforces, easy access to the Web to advertise their products, learn about competitors, and bid on jobs all over the world, and the availability of overnight delivery, they can now compete with SMEs in the United States in terms of cost, delivery, quality, service, technology, and all of the other requirements of integrated supply chains.

To respond to these converging challenges, U.S. small and medium-sized manufacturing enterprises must, at a minimum, take the following key steps:

- engage in meaningful strategic planning, not just budgeting
- increase their financial, managerial, and technological strengths
- add value to their products and integrate more closely with their customers
- integrate their own supply chains to reduce costs and improve performance

These responses will not, by themselves, ensure competitiveness, but they are essential for the successful participation of small and medium-sized manufacturing enterprises in modern integrated supply chains.

References

Agility Reports. 1997. Remaking the Customer-Supplier Relationships: Business Process Integration and the Agile Enterprise. Bethlehem, Pa.: Agility Forum.

Astronics Corporation. 1995. Annual Report. Buffalo, N.Y.: Astronics Corporation.

Astronics Corporation. 1998. Annual Report. Buffalo, N.Y.: Astronics Corporation.

Baba, M., D. Falkenburg, and D. Hill. 1996. Technology management and American culture: implications for business process redesign. Research-Technology Management 39(6): 44–54.

Bellman, G.M. 1993. Getting Things Done When You Are NOT in Charge. New York: Simon and Schuster.

Blanchard, D. 1999. Flow manufacturing pulls through. Evolving Enterprise 2(1): 14–19.

The Boeing Company. 1997. Subcontracting Manual. St. Louis, Mo: The Boeing Company.

Burke, J. 1999. Now what? Forbes ASAP Big Issue no. 4: 97–100.

Camp, R.C. 1995. Business Process Benchmarking: Finding and Implementing Best Practices. Milwaukee, Wis.: ASQC Quality Press.

Carr, K. 1998. Supply Chain Optimization. Presentation by Kevin Carr, National Institute of Standards and Technology, to the Committee on Supply Chain Integration, National Research Council, Washington, D.C., February 17, 1998.

Cohen, J. 1999. Taking care of business. The Industry Standards 13: 78–82.

Drucker, P.F. 1998. Management's new paradigms. Forbes 162(7): 152–177.

Dyer, J.H. 1996. Ideas at work: how Chrysler created American keiretsu. Harvard Business Review 74(4): 42–56.

Ferguson, T.W. 1999. The technology that won't die. Forbes 163(7): 56.

Goldratt, E.M. 1990. What Is This Thing Called Theory of Constraints and How Should It Be Implemented? Croton-on-Hudson, N.Y.: North River Press.

Grover, M.B. 1999. Land grab. Forbes 164(4): 90–92.

Hong, P. 1998. Surfing CATIA. Road and Track 49(11): 150–155.

Integrated Manufacturing Technology Roadmapping Project. 1998. Information Systems for the Manufacturing Enterprise. Oak Ridge, Tenn.: Oak Ridge Centers for Manufacturing Technology.

Kanter, R.M. 1998. Simultaneity. Forbes ASAP Big Issue 3: 219–220.

LaLonde, B. 1997. Where's the beef in supply chain management? Supply Chain Management Review 33(3): 9–10.

Lambert, D.M., M.C. Cooper, and J.D. Pagh. 1998. Supply chain management: implementation issues and research opportunities. International Journal of Logistics Management 9(2): 1–18.

Magretta, J. 1998. The power of virtual integration: an interview with Dell Computer's Michael Dell. Harvard Business Review 76(2): 72–84.

Manufacturing Engineering. 1999. Digital cars at GM. Manufacturing Engineering 122(3): 32.

Manufacturing News. 1998. U.S. manufacturers haven't yet mastered the supply chain management game. Manufacturing News 5(12): 5–6.

Murphy-Hoye, M. 1999. Supply Chain Integration in the Electronics/PC Industry. Viewgraphs used in a presentation to the Supply Chain World Conference and Exposition, Chicago, Illinois, April 26–28, 1999.

Owen, J.V. 1999. Engineers do the vision thing. Manufacturing Engineering 15(3): 96.

Royal, W. 1999. Death of salesmen. Industry Week 248(10): 59–60.

Tanzer, A. 1999. Warehouses that fly. Forbes 164(10): 120–124.

Velocci, A. 1998. Pursuit of six sigma emerges as industry trend. Aviation Week and Space Technology 149(20): 52–57.

Weber, C. 1997. Robert C. Byrd Institute: At the leading edge of manufacturing technology. Presentation by C. Weber, Robert C. Byrd Institute, to the Committee on Supply Chain Integration, National Research Council, Huntington, West Virginia, June 17, 1997.

Wreden, N. 1999. Supply chain management: no company is an island. Beyond Computing 8(4): 32.

Youtie, J., and P. Shapira. 1997. Manufacturing needs, practices, and performance in Georgia, 1994–1998. Atlanta: Georgia Manufacturing Extension Alliance. Also available on line at: *http://www.cherry.gatech.edu/mod/pubs/execsum.html.*

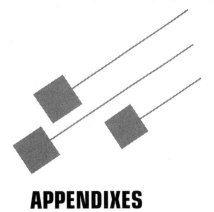

APPENDIXES

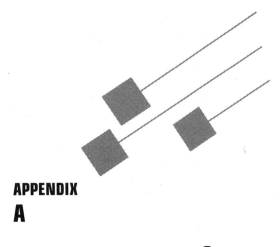

APPENDIX
A

Survey: Characteristics of Small Manufacturers

Appendix A contains a copy of the survey the committee sent to small and medium-sized manufacturing enterprises and a summary of the results.

NATIONAL RESEARCH COUNCIL
COMMISSION ON ENGINEERING AND TECHNICAL SYSTEMS

BOARD ON MANUFACTURING AND ENGINEERING DESIGN

Office Location:
Harris Building, Room 262
2001 Wisconsin Avenue, N.W.
Phone: (202) 334-3129
FAX: (202) 334-3718
rrusnak@nas.edu
Mailing Address:
2101 Constitution Avenue, NW
Washington, DC 20418

June 16, 1998

Dear Survey Participant:

On behalf of the National Research Council, I want to thank you for your participation. The objective of this survey is to determine what you think is important to satisfy your customers and to be a supplier for companies using modern supply chain management methods. This information will be used in conjunction with inputs from large companies to determine the most important attributes of good suppliers. Based on this analysis, the strategies will be developed for helping you to obtain needed capabilities.

We are not asking for your identity in the survey so you can be assured that your input is anonymous. In addition, we will only be using compiled data in our analysis, so individual results will not be revealed.

Inputs should be sent directly back to this office. Electronic response is strongly preferred (RRusnak@NAS.edu).

Again I want to thank you for your contribution. We anticipate that the ultimate results of this study will be used to assist you in becoming a first rate supplier and expanding your business.

Sincerely,

Robert M. Rusnak
Study Director

SURVEY
FACTORS FOR SUCCESS IN SUPPLY CHAINS

NATIONAL RESEARCH COUNCIL

Characteristics of Respondent

Industry   ———————— ———————— (SIC)

Sales   ————————

No. of employees   ————————
Primary product(s)   ————————

Build-to-print or other   ————————

% of sales from top three   ————————
customers   ————————
      ————————

List the SIC codes of top   ————————
three customers   ————————

Questions:

1. What percentage of your business transactions with your customer is done electronically?

2. To what extent do your top three customers share their future product and technology plans with you?

Insert number to reflect level:

1	2	3	4	5
never				always

Customer 1 ___
Customer 2 ___
Customer 3 ___

3. Do you have the following capabilities?

	(Now)	(Near Future)
SPC	_____	_____
CAD	_____	_____
CAM	_____	_____
MRP	_____	_____
ISO/QS	_____	_____
HAZMAT	_____	_____

4. How would you characterize your relationship with your top three customers?

Insert number to reflect level:

```
   1       2      3      4      5
adversarial                 full partner
```

Customer 1 ___
Customer 2 ___
Customer 3 ___

5. What critical factors or new capabilities would improve your success as a supplier?

Insert number to reflect level:

```
   1        2      3      4      5
not important              very important
```

Payment terms ____
Customer recognition program ____
Sharing of cost data ____
Sharing of objective performance data ____
Early involvement in customer product development ____
Production forecast ____
Financing ____

Others

_____

_____

_____

_____

SURVEY RESULTS

There were 99 completed questionnaires, one of which was from a very small enterprise whose owners and employees only participate in the company on a part-time basis. The data from this company was not included in the database. Hence, the total sample size is 98.

TABLE A-1 General Characteristics

Question	Number of Responses	Mean	Median	Standard Deviation
Annual sales ($ million)	82	34.9	7.7	111.3
Number of employees	95	226.7	75.0	514.8
Sales to top three customers as a percentage of total sales	86	42.4	34.3	25.8
Percentage of transactions with customers performed electronically	96	11.4	2.0	17.4
Extent to which customers share product and technology plans (1 [low] to 5 [high])	97	2.8	3.0	1.1

TABLE A-2 Capabilities of SMEs

Capability	Number of Respondents	Percentage with Capability Now	Percentage Planning to Develop Capability	Total (Percent)
SPC	98	55	12	67
CAD	98	74	1	75
CAM	98	47	11	58
MRP	98	43	16	59
ISO/QS	98	41	35	76
HAZMAT	98	48	7	55

TABLE A-3 Relations with Top Three Customers

Number of Responses	Mean	Median	Standard Deviation
98	3.7	3.7	0.6

Scale: 1 = adversarial; 5 = full partner

TABLE A-4 Factors That Would Improve Probability of Supplier Success

Factors	Number of Responses	Mean	Median	Standard Deviation
Improved payment terms	96	2.8	3.0	1.3
Better customer recognition programs	94	2.6	3.0	1.1
More extensive sharing of cost data	96	2.4	2.0	1.2
More extensive sharing of performance data	95	3.5	4.0	1.2
Earlier involvement in product development	95	4.1	5.0	1.1
More extensive sharing of production forecasts	94	3.8	4.0	1.2
Better financing	93	2.4	2.0	1.3

Scale: 1 = not important; 5 = important

TABLE A-5 General Characteristics of Subsamples

Question (averages)	Large SMEs	Small SMEs	Dispersed Customer Base	Concentrated Customer Base
Number of respondents	41	41	43	43
Annual sales (millions of dollars)	65.4	3.5	47.5	14.5
Number of employees	390.5	40.7	300.1	121.5
Sales from top three customers as a percentage of total sales	34.8	42.3	19.5	65.6
Percentage of transactions with customers performed electronically	10.9	13.2	7.5	15.4
Extent to which customers share product and technology plans (1 [no] to 5 [yes])	3.1	2.4	2.8	2.7

TABLE A-6 Capabilities of Large and Small SMEs

Capability	Percentage with Capability Now		Percentage Planning to Develop Capability	
	Large SMEs	Small SMEs	Large SMEs	Small SMEs
Number in sample	41	41	41	41
SPC	66	44	10	12
CAD	88	59	0	0
CAM	56	37	10	7
MRP	59	22	15	17
ISO/QS	54	22	27	44
HAZMAT	59	39	7	5

TABLE A-7 Capabilities of SMEs with Dispersed and Concentrated Customer Bases

Capability	Percentage with Capability Now		Percentage Planning to Develop Capability	
	Dispersed Customer Base	Concentrated Customer Base	Dispersed Customer Base	Concentrated Customer Base
Number in sample	43	43	43	43
SPC	44	63	14	9
CAD	72	70	0	2
CAM	49	47	12	7
MRP	42	47	19	14
ISO/QS	35	44	49	21
HAZMAT	47	49	5	9

TABLE A-8 Success Factors in Subsamples

Average importance of factors (based on scale of 1 to 5)	Large SMEs	Small SMEs	Dispersed Customer Base	Concentrated Customer Base
Number of respondents	41	41	43	43
Average relationship with top three customers	3.7	3.7	3.6	3.7
Payment terms	2.5	2.9	2.4	3.0
Customer recognition programs	2.4	2.6	2.5	2.5
Sharing of cost data	2.4	2.3	2.2	2.6
Sharing of performance data	3.4	3.3	3.3	3.3
Early involvement in product development	4.0	3.9	4.0	4.0
Sharing of production forecast	3.9	3.5	3.4	3.9
Financing	2.0	2.4	2.2	2.4

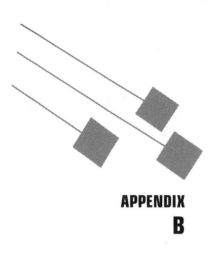

Capability Mapping

Capability mapping is a technique that can be used by supply chain participants to lay out, in an organized way, all of the critical functions, processes, and capabilities required to design, build, distribute, sell, and support the end product. The committee suggests that small and medium-sized manufacturing enterprises (SMEs) first map key individual processes and then superimpose them on one supply chain map. Maps of requirements can be overlaid with the actual capabilities of each participant to identify gaps (deficiencies in capabilities and capacities) systematically in the supply chain. These potential problem areas are candidates for careful data gathering, monitoring, and remedial action. Similar techniques can be used to create flow charts that map material flow requirements, capabilities, and deficiencies/gaps. Technological requirements, advances, and impacts can be mapped in a similar manner. Capability mapping can also be useful for systematically assessing candidates (both suppliers and customers) for participation in a supply chain, as well as for self-assessment by an SME.

Attempts to map, integrate, and manage all functions and process links are impractical, especially for SMEs. Some capabilities, functions, processes, and links are more important than others, and it is crucial to identify the most significant ones and prioritize the allocation of scarce resources accordingly. The following are examples of capability mapping parameters:

Capability requirements for product and process suppliers:

- demonstrated financial strength commensurate with the risk involved in becoming part of the supply chain
- ability to provide consistently high quality products and services
- effective and reliable process control systems throughout its own enterprise and throughout its own supply chains
- compliance with accepted industry standards for quality assessment and quality maintenance (e.g., IQ 9000)
- compliance with accepted industry standards for manufacturing enterprise performance (e.g., ISO 9000)
- compliance with applicable state and federal regulations
- validated process technologies appropriate for the products and services being supplied
- consistent on-time delivery
- ability to react with short lead times without incurring excessive costs or degrading quality
- ability to design and fabricate prototypes quickly and accurately to meet requirements for product realization programs
- use of appropriate data and information management systems that can effectively communicate with customers' and suppliers' systems and provide the information required to participate effectively in the supply chain.
- ability to produce the required products and/or services at competitive costs
- emphasis on employee learning and a well implemented training program

Unique capability requirements for product providers:

- ability to provide special insight into uses for their products and methods for adapting them to nonstandard applications
- product planning, design, and development capabilities that, although focused on creating and updating the supplier's own line of products, can be modified or adapted for nonstandard applications
- ability to provide after-sales support appropriate for dealing with unique or proprietary products.

Unique capability requirements for process suppliers:

- ability to guide customer product design decisions to reduce process costs and improve quality
- ability to provide process design capability to meet customer needs

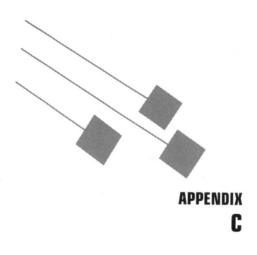

APPENDIX

C

Biographical Sketches of Committee Members

James Lardner (chair) retired as vice president for tractor and component operations at Deere & Company, where he implemented supply chain integration methods, including supplier participation in product design and integration of communication and computer tools. These techniques contributed to a reduction in product development cycle times of more than 60 percent. Mr. Lardner has served the National Research Council (NRC) as chair of the Manufacturing Studies Board, member of the Commission on Engineering and Technical Systems, and member of seven manufacturing committees. He has also been a member of five panels or committees for the National Academy of Engineering. Mr. Lardner is a fellow of the Society of Manufacturing Engineers and a current member of the boards of directors of two companies. He has extensive experience in implementing supply chain integration, as well as in manufacturing.

Steven J. Bomba is vice president for technology at Johnson Controls, Inc. He has participated in supply chain initiatives in several manufacturing divisions of Johnson Controls, both as a supplier and as a customer. He has been active in addressing the needs of small and medium-sized manufacturing supply chain participants in the Milwaukee area. Prior to joining Johnson Controls, he held a number of management positions, including vice president of advanced manufacturing technologies at Rockwell International. His expertise is in supplier operations and capabilities, as well as supply chain management. Dr. Bomba has served on several NRC committees, was a participant in the Japan-U.S. Manufacturing Research

Exchange, and was a member of the Panel on Equipment Reliability and Productivity. He is a past member of the NRC Manufacturing Studies Board.

John A. Clendenin, recently manager of new business ventures at Xerox Corporation, is now a professor at Harvard University. Previously, Mr. Clendenin was a manager of integrated supply chain strategies and business processes and held several other positions in which he was responsible for supply chain development. He has delivered keynote addresses at conferences on supply chain logistics and distribution and has developed workshops for business professionals. Mr. Clendenin was chair of the Best Practices in Supply Chain Management Conference in 1997. He has an extensive background in supply chain integration and operations.

Gerald E. Jenks is department head, supplier management, at The Boeing Company in St. Louis. In his current capacity, he is responsible for the development and implementation of supplier management methods and processes. He previously was manager, F-15 supplier management, where he was responsible for all supply chain activities and implemented a new supplier management approach. Prior to that, he was manager of a Harpoon Program subcontract, where he initiated methods for improving the quality and schedule performance of suppliers. He has extensive experience in supply chain operations, particularly with suppliers of defense products.

John J. Klim, Jr., is the president of D&E Industries, Huntington, West Virginia, a successful small business that makes forged and machined parts for the transportation, construction, and mining industries. He is a past chairman of the Advisory Board of the West Virginia District of the U.S. Small Business Administration and was a 1987 delegate to the White House Conference on Small Business. He is vice president of the Huntington Regional Chamber of Commerce and a member of the Board of Directors of the West Virginia Chamber of Commerce.

Edward Kwiatkowski is president and chief executive officer of Supply America Corporation, a nonprofit company that provides information, decision support, and implementation assistance to supply chains adapting advanced manufacturing technologies and business practices. Mr. Kwiatkowski was previously vice president of CAMP, Inc., where he was responsible for management of the Great Lakes Manufacturing Technology Center and the Electronics Resource Center. Both of these centers worked with small manufacturers to improve their manufacturing products and processes. He also served on the Board of Directors for the Modernization Forum and the National Coalition for Advanced Manufacturing.

Hau Lee is the Klein Perkins, Mayfield, and Sequoia Capital Professor and deputy chairman of the Department of Industrial Engineering and Engineering Management at the Stanford University Graduate School of Business. He is the founder and current director of the Stanford Global Supply Chain Management Forum, an industry-academic consortium to advance the theory and practice of global supply chain management. Dr. Lee is a consultant on global supply chain management for several Fortune 100 companies. He is on the editorial board of many international journals, including *Supply Chain Management Review*, and has published 42 papers on manufacturing operations and supply chain systems.

Charles W. Lillie is assistant vice president at Science Applications International Corporation, where he has held several positions related to the development and management of computer resources, including director of management information systems. In the past three years he established a division to focus on electronic commerce and Worldwide Web technologies for communication between prime manufacturers and their suppliers. Dr. Lillie recently chaired an international conference on software reuse and has chaired six annual workshops since 1992.

Mary C. Murphy-Hoye is program manager for strategic programs at Intel Corporation where she is responsible for supplier development and supply chain operations. She has developed and implemented a corporate-wide network for interacting with suppliers. In earlier assignments, she implemented a global inventory management system and a supply chain management system. Ms. Murphy-Hoye is currently a member of the Electronic Company Supply Chain Integration and Systems Consortium, an association of leading companies in the electronics industry.

James R. Myers is a partner in the law firm Kilpatrick Stockton LLP, specializing in intellectual property. He is currently general counsel for the National Initiative for Supply Chain Integration, which is committed to developing advanced technologies for improving supply chain performance. Members of this organization include companies in the forefront of supply chain integration, such as DaimlerChrysler, Proctor & Gamble, Honda, Harley Davidson, Trane, and Deere. Mr. Myers also participated in the National Institute of Standards and Technology Manufacturing Extension Partnership. His expertise is in intellectual property issues, particularly in the context of integrated supply chains.

James B. Rice, Jr., is director of the Integrated Supply Chain Management Program at the Massachusetts Institute of Technology (MIT), an industry-academia collaborative research program focused on supply chain integration. Mr. Rice also teaches the "Supply Chain Context" course in the Master of Engineering in Logistics degree program at MIT and conducts research on the design of supply chain management organizations and systems. He recently developed a strategic framework, taxonomy, and recommended practices for managing the horizontal processes of supply chains. Prior to his current appointment at MIT, Mr. Rice held several positions in manufacturing and logistics at The Procter & Gamble Company. He is currently a member of the Integrated Advisory Board for the KLICT supply chain research initiative in the Netherlands, the editorial board of *Supply Chain Management Review*, and the Board of Directors of the New England Chapter of the Council of Logistics Management.

Oliver Williamson is professor of business administration, professor of economics, and professor of law at the University of California, Berkeley. He has published widely on economic efficiency and the design of corporate and legal systems as related to the structuring of prime contractors and suppliers in integrated supply chains. In addition to his membership in the National Academy of Sciences, Dr. Williamson is a fellow of the American Academy of Political and Social Science, the American Academy of Arts and Sciences, and the Econometrics Society.

Thomas Young is a retired president and chief operating officer of Martin Marietta Corporation and executive vice president of Lockheed Martin. He previously headed Martin Marietta Corporation's Electronics and Missiles Group, where supply chain management was a major facet of operations. Mr. Young has served on the NRC Committee on Space Technology Needs for the Future and was chair of the Committee on the International Space Station.

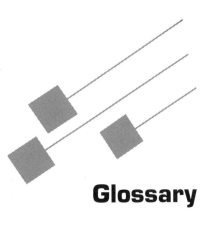

Glossary

Computer-aided design (CAD). A combination of computer software and hardware used in conjunction with computer graphics to enable engineers and designers to create, manipulate, and change designs without conventional paper drafting.

Computer-aided manufacturing (CAM). The use of computers to control and monitor manufacturing elements, such as robots, computer numerical control machines, storage and retrieval systems, and automated guided vehicles. At the lowest level, CAM includes programmable machines controlled by a centralized computer. At the highest level, large-scale systems integration includes control and supervisory systems.

Computer-integrated manufacturing (CIM). The integration of computer systems in a manufacturing facility. Integration may extend beyond the factory into the facilities of suppliers and customers. CIM integrates systems that handle everything from ordering to shipment of the final product, including accounting, finance, management, engineering, and manufacturing. The scope of CAM is generally limited to the factory floor, but CIM generally extends beyond the factory floor.

Concurrent engineering (CE). An approach in which product design, process development, and manufacturing preparations are carried out simultaneously.

e-Business. Using the capabilities of Internet technology, including turning raw information and data into actionable intelligence, to conduct business electronically.

e-Commerce. Buying, selling, and exchanging information electronically.

Extranet. A collaborative network that uses Internet technology to link businesses with their supply chains and provides a degree of security and privacy from competitors.

Firewall. A combination of hardware and software designed to make a Web site secure.

Hypertext markup language (HTML). A hypertext document format used on the Worldwide Web. Tags and directive information are embedded in the document to delimit text and indicate special instructions for processing it.

Integrated supply chain. An association of customers and suppliers who, using management techniques, work together to optimize their collective performance in the creation, distribution, and support of an end product.

Internet. A worldwide collection of servers and networks that allow users access to information and applications outside of the company firewall.

Intranet. A secured network of Web pages and applications that can be accessed by anyone within a company firewall.

Just-in-time (JIT) manufacturing. An approach in which goods and services are produced only when needed for the next manufacturing step, rather than being stockpiled in advance.

Local area network (LAN). A communication system within a facility; the backbone of a communication system that connects various devices in a factory to a control center. The LAN, through the control center, allows devices, such as computers, bar code readers, programmable controllers, and CNC machines, to communicate with each other for control and exchange of information.

Materials requirements planning (MRP or MRP I). A scheduling technique for establishing and maintaining valid due dates and priorities for orders based on bills of material, inventory, order data, and the master production schedule.

Manufacturing resource planning (MRP II). A direct outgrowth and extension of closed-loop MRP I through the integration of business plans, purchase commitment reports, sales objectives, manufacturing capabilities, and cash-flow constraints.

Modeling and simulation. The application of a rigorous, structural methodology to create and validate a physical, mathematical, or otherwise logical representation of a system, entity, phenomenon, or process for making managerial or technical decisions.

Original equipment manufacturer (OEM). A manufacturer that builds products for end users rather than components for use in other products.

Outsourcing. The procurement of goods and services from suppliers outside of the corporation.

Partners. Companies that agree to work together, often for a specific period of time or to achieve specific objectives, and share the risks and rewards of their relationships.

Partnership. An agreement between two companies, often formalized in a contract.

Small and medium-sized manufacturing enterprise (SME). A manufacturing company with fewer than 500 employees.

Statistical process control (SPC). The use of statistical techniques for monitoring and controlling the quality of a process and its output over time. SPC can be used to reduce variability in processes and output quality.

Supply chain. An association of customers and suppliers who, working together yet in their own best interests, buy, convert, distribute, and sell goods and services among themselves resulting in the creation of a specific end product.

Supply chain management. The integration of key business processes, from end user through original suppliers, that provide products, services, and information that add value for customers and other stakeholders.

Transparency. The extent that participants are aware of activities throughout the supply chain.

Virtual enterprise. An opportunity-driven partnership or association of enterprises with shared customer loyalties designed to share infrastructure, research and development, risks, and costs and to link complementary functions.

Bibliography

SUPPLY CHAIN CONCEPTS

Theory and Concepts

Bovet, D.S., and Y. Sheffi. 1998. The brave new world of supply chain management. Supply Chain Management Review 34(4): 14–22.

Bowman, R.J. 1997. Link by link: global supply chains. Distribution 96(8): 88–90.

Burgess, R. 1998. Avoiding supply chain management failure: lessons from business process re-engineering. International Journal of Logistics Management 9(1): 15–23.

Copacino, W.C. 1998. Copacino on strategy: get the complete supply chain picture. Logistics Management and Distribution Report 37(11): 45.

Fox, M.L., and J.L. Holmes. 1998. A model for market leadership. Supply Chain Management Review 34(2): 54–61.

Krause, D., and R. Handfield. 1999. Developing a world class supply base. Tempe, Ariz.: National Association of Purchasing Managers Center for Advanced Purchasing Studies.

La Londe, B.J. 1999. Executing in the red zone. Supply Chain Management Review 35(2): 7–9.

Mariotti, J.L. 1999. The trust factor in supply chain management. Supply Chain Management Review 35(2): 70–77.

Narus, J.A., and J.C. Anderson. 1996. Rethinking distribution: adaptive channels. Harvard Business Review 74(4): 112–120.

Schlegel, G.L. 1999. Supply chain optimization: a practitioner's perspective. Supply Chain Management Review 35(1): 50–57.

Approaches and Trends

Atkinson, H. 1999. Sun is rising on plan to remove bar between customers, suppliers. Journal of Commerce, 419(29, 379): 1–14.

Fine, C. 1999. The primacy of chains. Supply Chain Management Review 35(2): 79–88.

Guzak, R.J., and D.M. Hill. 1998. Operations franchise: leveraging your supply chain for profitable growth. Supply Chain Management Review 34(2): 62–69.

Hutchinson, B., and J.G. Welty. 1998. Global trends in the consumer markets. Supply Chain Management Review 34(4): 58–66.

Laseter, T.M. 1999. Integrating the supply web. Supply Chain Management Review 35(1): 87–94.

Mentzer, J.T., and C.C. Bienstock. 1998. The seven principles of sales-forecasting systems. Supply Chain Management Review 34(4): 76–83.

Schwalbe, R.J. 1998. SMART 2001: supply chain management, Siemens style. Supply Chain Management Review 34(4): 69–75.

SUPPLY CHAIN INITIATIVES

Financial

Cooke, J.A. 1998. Panning for gold. Logistics Management and Distribution Report 37(11): 59–62.

Malone, R. 1998. Balancing inventory with customer service. Inbound Logistics 18(8): 10–11.

Inventory

Quinn, F.J. 1998. Balancing demand and supply. Logistics Management and Distribution Report 37(10): 67.

Logistics

Huppertz, P. 1999. Market changes require new supply chain thinking. Transportation & Distribution 40(3): 70–74.

Manufacturing

Adaptable manufacturing: agility to cost-effectively produce on demand. 1998. Modern Materials Handling 53(6): 14–16.

Postponement

van Hoek, R.I. 1998. Reconfiguring the supply chain to implement postponed manufacturing. The International Journal of Logistics Management 9(1): 95–110.

Postponement systems: waiting to the last minute is a virtue. 1998 Modern Materials Handling 53(6): 27–28.

Procurement

Anderson, M.G., and P.B. Katz. 1998. Strategic sourcing. International Journal of Logistics Management 9(1): 1–13.

Handfield, R.B., and D.R. Krause. 1999. Think globally, source locally. Supply Chain Management Review 35(1): 36–46.

Hoffman, K.C. 1998. Automakers push for ever-closer collaboration with suppliers. Global Sites and Logistics 2(8): 38–42.

Warehousing and Distribution Centers

Bowman, R.J. 1998. It's not your father's warehouse. Warehousing Management 5(5): 32–33.

Third-Party Providers

Menon, M.K., M.A. McGinnis, and K.B. Ackerman. 1998. Selection criteria for providers of third-party logistics services: an exploratory study. Journal of Business Logistics 19(1): 121–137.

Thomas, J. 1999. Chain reaction. Logistics Management and Distribution Report 38(1): 55.

Reverse Logistics

Buxbaum, P. 1998. The reverse logistiX files. Inbound Logistics 18(9): 62–67.

Marien, E.J. 1998. Reverse logistics as competitive strategy. Supply Chain Management Review 34(2): 43–52.

SUPPLY CHAIN IMPLEMENTATION

Benchmarking and Metrics

Dawe, R.L. 1998. Mobilizing for global excellence. Supply Chain Management Review 34(2): 11–13.

Jennings, B.D. 1998. Supply chain economics making your shots count. Logistics, Winter/Spring newsletter, pp. 2–5. Mercer Management Consultants, 2300 N. St N.W., Washington, D.C. 20037.

Cost Analysis

Braithwaite, A., and E. Samakh. 1998. The cost-to-serve method. International Journal of Business Logistics 9(1): 69–84.

La Londe, B.J. 1998. The costs of "functional shiftability." Supply Chain Management Review 34(1): 9–10.

e-Commerce

Gross, N. 1998. The supply chain: leapfrogging a few links. Business Week 3583: 140–142.

Mottley, R. 1998. Spinning supply chains via the Internet. American Shipper 40(11): 26–28.

Information Technology Integration and Enterprise Resource Planning

Appleton, E.L. 1997. Supply chain brain. CFO 13(7): 51–54.

Bundy, W. 1999. Leveraging technology for speed and reliability. Supply Chain Management Review 35(2): 62–69.

Buxbaum, P. 1998. Technology tightens links in supply chain. Transport Topics 3308: 12-14.

CASE STUDIES

Industry

Brunell, T. 1999. Managing a multicompany supply chain. Supply Chain Management Review 35(2): 45–52.

Simison, R.L. 1997. New data illustrates reshaping of auto parts business: firm's makeover reflects car industry's assigning more assembly to suppliers. Wall Street Journal, September 2, 1997.

Stein T., and J. Sweat. 1998. Killer supply chains. Information Week 708: 36–46.

Retail and Wholesale

Hoffman, K.C. 1999. Elizabeth Arden's supply chain gets dramatic facelift. Global Sites and Logistics 3(1): 20–28.
Lampe, J., and R.W. Gray. 1998. The Bridgestone/Firestone perspective: betting on the supply chain. Supply Chain Management Review 34(2): 24–30.
Reuland, T. 1999. Thomson Consumer Electronics: listening to the customer's voice. Supply Chain Management Review 35(1): 28–34.

Global Supply Chain

Fites, D.V. 1996. Make your dealers your partners. Harvard Business Review 74(2): 84–95.
Harrington, L.H. 1999. The high tech sector: meeting supply challenges at warp speed. Transportation and Distribution 40(3): 49–53.
Hoyt, J.B. 1998. Lessons learned on a supply chain journey. Supply Chain Management Review 34(4): 84–92.
Margerita, J. 1998. Fast, global, and entrepreneurial: supply chain management, Hong Kong Style. An interview with Victor Fung. Harvard Business Review 76(5): 103–114.
McIntyre, K., H.A. Smith, A. Henham, and J. Pretlove. 1998. Logistics performance measurement and greening supply chains: diverging mindsets. International Journal of Logistics Management 9(1): 57–68.
Scharlacken, J.W. 1998. The seven pillars of global supply chain planning. Supply Chain Management Review 34(2): 32–40.

Manufacturing

John, C.G., and M. Willis. 1998. Supply chain re-engineering at Anheuser-Busch. Supply Chain Management Review 34(4): 28–36.

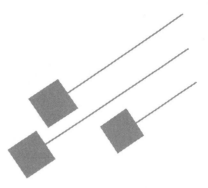

Index